D1435415

Microbiology

Disclaimer

Adult supervision is required when working on these projects. All the projects within this book assume your teacher or some other knowledgeable adult advisor will be assisting you throughout the project. No responsibility is implied or taken for anyone who sustains injuries as a result of using the materials or ideas, or performing the procedure put forth in this book.

Use proper equipment (gloves, forceps, safety glasses) and take other safety precautions. Read and follow the manufacturer's instructions when using equipment and chemicals, dry ice, boiling water, flames, or any heating elements with extra care. Wash hands after project work is done. Taste nothing. Tie up loose hair and clothing. Follow step-by-step procedures and avoid shortcuts. Never work alone. Additional safety precautions are mentioned throughout the text and in the section "Safety in the Microbiology Laboratory." If you use common sense and make safety a first consideration, you will create safe, fun, educational, and rewarding projects.

Microbiology

High-School
Science Fair Experiments

H. Steven Dashefsky

TAB Books

Division of McGraw-Hill, Inc.

New York   San Francisco   Washington, D.C.   Auckland   Bogotá
Caracas   Lisbon   London   Madrid   Mexico City   Milan
Montreal   New Delhi   San Juan   Singapore
Sydney   Tokyo   Toronto

pbk   1 2 3 4 5 6 7 8 9 DOC/DOC 9 9 8 7 6 5 4
hc   1 2 3 4 5 6 7 8 9 DOC/DOC 9 9 8 7 6 5 4

Library of Congress Cataloging-in-Publication Data
Dashefsky, H. Steven.
    Microbiology : high-school science fair experiments / by H. Steven
Dashefsky.
        p. cm.
    Includes bibliographical references (p. 163) and index.
    ISBN 0-07-015663-8 (H)   ISBN 0-07-015664-6 (P)
    1. Microbiology—Laboratory manuals. 2. Biology projects.
I. Title.
QR63.D35   1994
576'.078—dc20   94-23117
                                                          CIP

Acquisitions editor: Kimberly Tabor
Editorial team: Joanne M. Slike, Executive Editor
              David M. McCandless, Managing Editor
              Marianne Krcma, Book Editor
Production team: Katherine G. Brown, Director
              Susan E. Hansford, Coding
              Stephanie A. Myers, Computer Artist
              Wanda S. Ditch, Desktop Operator
              Joann Woy, Indexer
              Linda L. King, Proofreading   0156646
Design team: Jaclyn J. Boone, Designer   GEN1

Acknowledgments

Many of the experiments in this book were adapted from original International Science and Engineering Fair (ISEF) projects. (All of the projects were edited, and many were modified for this book.) I want to thank the following young scientists for their outstanding projects and wish them the best of luck in their future scientific endeavors:

- Jessica Sweeney for "Oysters, Hot Sauce, and Bacteria: Does Hot Pepper Sauce Kill Bacteria in Raw Oysters?" (chapter 4)
- Neelu Jain for "Trees and Bacteria: Can Extracts from Tree Bark Inhibit the Growth of Bacteria?" (chapter 6)
- Anita Jo Williams for "Common Spice and Microbes: Can the Vapors from Common Spices Control Microbe Growth?" (chapter 10)
- Heather Ewald for "Garlic Juice and Microbes: Does Garlic Juice Control Bacteria?" (chapter 11)
- John Howard for "Hot and Cold Garlic: Does Heating Garlic Cause it to Lose its Ability to Kill Bacteria?" (chapter 12)
- Carrie Kennemer for "Bottled Water and Microbes: How Pure Are Different Brands of Bottled Water?" (chapter 13)
- Jayme Steig for "Resistant Bacteria: Can Bacteria Become Resistant to Household Disinfectants?" (chapter 17)
- Karmen Hepper for "Electromagnetic Fields and Soil Organisms: Does Electromagnetic Radiation Affect Nematodes?" (chapter 20)
- Emil King for "Electromagnetic Fields and Aquatic Organisms: Does EMR Affect Brine Shrimp?" (chapter 23)

Contents

**A word about
safety & supervision** *x*

How to use this book *xiii*

Part 1
Before you begin

**1 An introduction
to microbiology** *3*
Terminology **4**
Sterile technique **5**
Microscopes **12**
Incubation **17**

**2 An introduction
to scientific research** *20*
The scientific method **21**
Building on past projects **23**

3 Getting started *24*
Use this book **24**
Other sources **25**
Talk to specialists in the field **25**
Before you begin **26**
Performing the experiment **26**

Part 2
Controlling & using microbes

   4 Oysters, hot sauce, & bacteria   *33*

   5 Microbial pest control   *38*

   6 Trees & bacteria   *44*

Part 3
Microbes on the move

   7 Insects & microbes   *51*

   8 Spreading germs   *55*

   9 Microbes in the wind   *61*

Part 4
Garlic, health, & microbes

10 Common spices & microbes   *69*

11 Garlic juice & microbes   *73*

12 Hot & cold garlic   *79*

Part 5
Microbes: they're everywhere

13 Bottled water & microbes   *87*

14 Molds in the home   *93*

15 Microbes on your body   *99*

16 Soils' microbes   *104*

Part 6

Controlling microbes with antimicrobial products

17 **Resistant bacteria** *111*

18 **Biofilms & biofouling** *118*

19 **Microbes in the home** *123*

Part 7

Microbes & humans

20 **Electromagnetic fields & soil organisms** *131*

21 **Buffers** *137*

22 **Gram stains** *141*

23 **Electromagnetic fields & aquatic organisms** *146*

Appendices

A **Using metrics** *154*

B **Addresses** *156*

   Glossary *158*

   Bibliography *163*

   Index *165*

   About the author *175*

A word about safety & supervision

Note: All the experiments in this book require an adult sponsor to ensure the student's safety and the safety of others. This sponsor should be your teacher or some other adult advisor who is knowledgeable about microbiology laboratory procedures, including safety and disposal techniques.

Science Service, Inc. is an organization that sets science fair rules, regulations, and safety guidelines. It also holds the International Science and Engineering Fairs (ISEF). This book recommends and assumes that students performing projects in this book follow ISEF guidelines as they pertain to adult supervision. ISEF guidelines state that students undertaking a science fair project have an adult sponsor assigned to them.

The adult sponsor is described by the ISEF guidelines as a teacher, parent, professor, or scientist in whose lab the student is working. (For the purpose of this book, this will usually be the student's teacher.) This person must have a solid background in science and be in close contact with the student throughout the project. The adult sponsor is responsible for the safety of the student while conducting the research, including the handling of all equipment, chemicals, and organisms.

The sponsor must be familiar with regulations and commonly approved practices that govern chemical and equipment usage; experimental techniques; the use of laboratory animals, cultures, and microorganisms; and proper disposal techniques. If the adult sponsor is not qualified to handle all of these responsibilities, the sponsor must assign another adult who can fulfill these responsibilities. Most science fairs require that appropriate forms identifying the adult sponsor and his or her qualifications be filled out before proceeding with a project.

The sponsor is responsible for reviewing the student's research plan, as described later in this book, and making sure the experimentation is done within local, federal, and ISEF (or other appropriate governing body) guidelines.

Both the student and the adult sponsor should read the entire project before it is begun. The adult should determine which portions of the experiment the student can perform without supervision and which portions will require supervision. In addition, a ⚠ throughout the text indicates procedures that are potentially hazardous.

For a copy of the ISEF's rules and regulations booklet, contact

Science Service, Inc.
1719 N Street NW
Washington, DC 20036
(202) 785-2255

The booklet includes a checklist for the adult sponsor, approval forms, and valuable information on all aspects of participating in a science fair. Everyone should have this information before competing in a science fair.

Safety in the microbiology laboratory

Both students and sponsors should always be conscious of possible contaminants, especially on your hands. Wear protective clothing including goggles, lab coats, and gloves. Clean up your work surface, especially if you had a spill of any kind. Wash your hands when the work is done. Never eat or drink in a laboratory. If you contaminate your clothing, remove it and wash it.

Some of the experiments in this book produce fungal (mold) growths. Fungi spores are easily carried in the air. If possible, do not open plates with mold growths. If you must, open them slowly and carefully, in a room with very little air movement.

Some microbes are *pathogenic* (disease-causing). None of the experiments in this book uses pathogenic microbes, but when you isolate microbes in an experiment, it is possible that you might culture some pathogens. Therefore, handle all cultures as if they were pathogenic. It is extremely important that you safely dispose of all cultures. The best way to dispose of cultures is to autoclave them before disposal to be sure all the microbes are dead and will not contaminate anything.

If you have cultures on agar plates (discussed later in this book), seal the plates with adhesive tape, wrap the plates in a biohazard bag and have your sponsor handle their proper disposal.

How to use this book

There are two ways to use this book. If you are new to science fair projects and feel that you need a great deal of technical guidance, you can use the projects as explained in this book with little or no adjustment. These are good, solid (and in many cases award-winning), science fair projects.

However, don't be afraid to use these projects as models for your own. Suggestions in the "Going further" and "Suggested research" sections of each project help you do this. Each project in this book provides a core experiment with many suggestions about how to expand its scope or adjust its focus.

The projects in this book

The projects in this book have been grouped into six parts:
- Controlling microbes and using microbes to control other organisms
- Microbes on the move
- Garlic, health, and microbes
- They're everywhere
- Trying to control microbes with antimicrobials
- Microbes and man

An introduction at the beginning of each part briefly describes the projects in that part of the book. Each project has the following sections:
- Background
- Project overview
- Materials list
- Procedures

- Analysis
- Going further
- Suggested research

Background

The "Background" section provides general information about the topic. It offers a frame of reference so you can see the importance of the topic and why research is necessary to advance our understanding of it. This section could be considered the initial step in your literature search (more about the literature search later in this chapter). Although a small step, reading this section is enough to see if the subject piques your interest.

Project overview

If the "Background" section of a particular project interests you, read the "Project overview" section, which describes the purpose of the project. This section explains the problems that exist and poses questions that the experiment is intended to solve. You can use these questions to formulate your hypothesis. Be sure to discuss this section, as well as the next, with your sponsor to see if you can realistically meet the requirements imposed.

Materials list

The "Materials list" section gives everything needed to perform the experiment. Be sure you have access to or can get everything before beginning. Some pieces of equipment are expensive; check with your teacher to see if all the equipment is available in your school, or can be borrowed from elsewhere. Be sure your budget can handle anything that must be purchased. (A list of scientific supply houses is provided at the back of the book.)

Although most people don't think of research scientists as being particularly good with a hammer and nails, they must be. Building a device or experimental workstation often involves many trips to the hardware store for supplies, a little sweat on the brow, and a lot of ingenuity.

Most living organisms, such as bacterial cultures, must be ordered from the supply houses. Others, such as insects, can be ordered, purchased locally, or caught depending on the project, your location, and the time of year. If you are using live organisms, work with your sponsor to be sure you adhere to all science fair regulations and standard biological research practices. Before beginning, discuss with your sponsor the proper way to dispose of any hazardous materials, chemicals, or cultures.

Procedures

The "Procedure" section gives step-by-step instructions on how to perform the experiment and suggestions on how to collect data. Be sure to read through this section with your sponsor before undertaking the project. Illustrations are often used to clarify procedures.

Analysis

The "Analysis" section doesn't draw conclusions for you. Instead, its questions help you analyze and interpret the data so you can come to your own conclusions. In many cases, empty tables and charts are provided for you to begin your data collection. You should convert as much of your raw data as possible into line graphs, bar graphs, or pie charts.

Some experiments might require statistical analysis to determine if there are significant differences between the experimental groups and the control group. Check with your sponsor to see if you should perform statistical analysis for your project and if so, what kind. The appendix called "Further reading" at the back of this book lists books that will help you analyze your data.

Going further

The "Going further" section is a vital part of every project. It lists ways for you to continue researching the topic beyond the original experiment. Although these suggestions can be followed as-is, more importantly, they might spark your imagination to think of some new twist or angle to take while performing the project. These suggestions might show ways to more thoroughly cover the subject, or show you how to broaden the scope of the project. The best way to ensure an interesting and fully developed project is to include one or more of the suggestions from the "Going further" section, or an idea of your own that was inspired from this section.

Suggested research

The "Suggested research" section proposes new directions to follow while researching the project. It might suggest what to read, or organizations, companies, and other sources to contact. Using these additional resources might turn your project into a winner.

Part 1

Before you begin

Before delving into any science experiment, you need to understand three things: the terminology used, the methodology required, and the suitability of the experiment to your own situation. The following three chapters examine these elements.

1

An introduction to microbiology

Microorganisms are found all over the world. They are in the soil, in the air, at the bottom of the sea, in hot springs, and almost every other place on our planet. A teaspoon of rich soil can contain billions of bacteria, millions of fungi, and hundreds of thousands of algae and protozoans.

Microorganisms, also simply called *microbes*, are usually divided into five major groups, as shown in Fig. 1-1:

- Bacteria
- Algae
- Fungi
- Protists
- Viruses

Some of these organisms are *autotrophs*, meaning they make their own food during photosynthesis just as green plants do. Others are *heterotrophs*, meaning they must consume their food.

Some microbes are predators that attack and devour other microbes. Others are parasites living in hosts that may be plants or animals. Still others are *saprophytes* (decomposers) that feed on dead, decaying organisms. These microbes play an important role in helping to recycle nutrients so they can be used by another generation of living plants.

In the strictest sense, only organisms belonging to one of the five groups just mentioned are considered microbes. For the sake of this book, however, we will use a broader definition of *microbe* to include any organisms requiring a microscope to study.

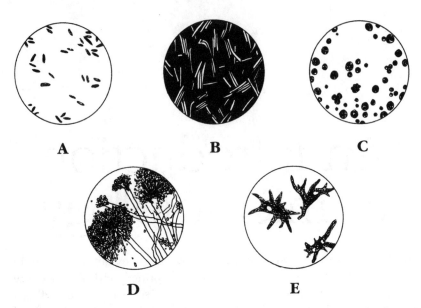

1-1 *Microbes include* (a) *viruses,* (b) *bacteria,* (c) *algae,* (d) *fungi, and* (e) *protists. (Not to scale.)*

Terminology

The small size of microorganisms poses problems for scientists trying to study them. One bacterium looks pretty much like another. Cells are often too small to study as individuals, so it is necessary to study a group of them together, called a *colony*. When you grow these colonies in an artificial environment such as a petri dish, they are called *cultures*. A *pure culture* of microbes contains only one type of microbe and has not been contaminated with any other type of microbe.

To create pure cultures, scientists use sterile techniques, also called *aspetic techniques*. These techniques include methods to sterilize glassware as well as particular pieces of equipment such as inoculating loops and spreading rods. Sterile techniques also include requiring a sterile material for the microbes to grow on called a *growth media*.

The growth media must nourish the microbes and provide a solid surface on which they can live and reproduce. *Agar*, a compound extracted from algae, gels to a solid after boiling and is therefore perfect for growing (culturing) colonies of microbes. Culturing pure colonies on a solid surface lets microbiologists see characteristics of the microbe such as the shape, size, and color of colonies.

Sources of microbes

Since individual microbe cells are too small to handle, you will deal with cultures. The microbe cultures are grown or purchased in vials with screw caps or petri dishes with lids that contain the culture media.

If a culture is purchased, it might come in vials containing a broth or containing solid nutrient agar. These vials with agar are called *agar slants*. Alternatively, the culture might come growing in a petri dish that contains a nutrient agar as the culture media. These petri dishes are often simply called *plates*.

Culture media

When you purchase a culture, whether it be in vials or plates, the microbes are growing on some form of media. When you grow your own cultures, they too must be grown on a media. This media supplies all the nutrients required by the microbes to survive and gives them a place to reproduce.

Most of the experiments in this book require you to grow pure cultures. This means that the media must be sterile. If the media is sterile, there won't be any microbes already on it before you introduce the microbes you want to study. Sterile culture media can be purchased from a scientific supply house, or you can sterilize a mix yourself as discussed later in this chapter.

There are probably as many recipes for microorganism culture media as there are microbiologists. A few standard recipes for culture media that many microbes can grow on are used throughout this book. "Nutrient agar," "blood agar," and "tryptic soy agar" are three examples. Some projects, however, require special media; such requirements are listed in the "Materials list" section of each project. You can purchase all of these media from the scientific supply houses listed in appendix B at the back of this book.

Sterile technique

While performing most of the projects in this book, it is important to avoid contaminating your work. You avoid contamination by keeping unwanted microbes out of your experiment. The only way to prevent these unwanted microbes from ruining your project is to use a good sterile, or aspetic, technique. Sterile technique is described here, but be sure to get your sponsor to demonstrate these techniques for you when they are needed for your project.

Sterile technique is used to transfer (*inoculate*) microbes from their source to a culture media without allowing any unwanted microbes to contaminate the media. This means using equipment, supplies, and media that are sterile (free of microbes). All glassware and utensils must be sterilized before they touch the culture, otherwise the culture can become contaminated with whatever microbes are on the utensils. (See the next section for details on sterilizing glassware and utensils.)

Since microbes can enter a culture from the air, sterile technique also includes keeping culture vials and plates closed as much as possible. All opening and closing of sterile containers should be kept to a minimum and be done in a room with very little air movement. When opening a petri plate, keep the lid over the bottom while inoculating the plate and quickly put it back in place, as shown in Fig. 1-2. Also, don't wave sterilized utensils such as inoculation loops, spreading rods, and pipettes around in the air while being used, since they could become contaminated. (Proper sterilization of these utensils is described in the next section of this chapter.)

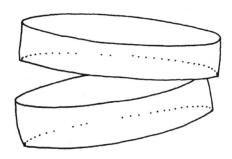

1-2
To maintain sterile technique, the lid of the petri dish should be replaced as quickly as possible.

Sterile technique means never placing a sterile item on an unclean countertop, a surface that usually contains more microbes than the air. Always keep your workspace clean. The workspace surface should be disinfected with a bleach and water solution (one part bleach to 100 parts water).

The most important thing about sterile technique is to keep yourself free from the microbes you are working with. Always wear protective clothing, including lab gloves and goggles. Always wash your hands when the work is done. This is especially important if you are working with unknown microbes. Since some of the microbes found in our environment might be *pathogenic* (disease-causing) you must take extra care.

Caution! Always consult with your teacher or advisor about proper handling and disposal of all microbes purchased or cultured for the projects in this book.

Sterilizing glassware & media

The best way to sterilize glassware and media is with an *autoclave*, shown in Fig. 1-3, which uses steam under pressure to kill all organisms. If you don't have an autoclave available, you can still sterilize glassware for the projects in this book by following these instructions:

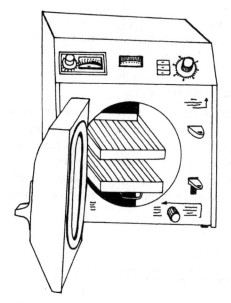

1-3
The best way to sterilize glassware and media is with an autoclave, which uses steam under pressure.

Start with clean glassware (jars) that have been well-rinsed with soapy water. Then, sterilize the glassware by filling them halfway with water and placing them in a 4-quart pot that has two or three inches of water in it. The water in the jars must be below the water line in the large pot. Place the jar lids into the large pot as well, and boil everything for 30 minutes.

When complete, empty the mason jars of the hot water by picking them up with sterile tongs and dumping out the water. However, place these empty jars back into the hot water until you are ready to use them. Also, remove the lids from the pot and put them to the side on a sterile surface to dry.

The empty glassware will remain sterile in the warm water for about 30 minutes because the rising hot air from the water in the pot

prevents any microbes from settling on the glassware. This sterilization procedure will not kill bacterial spores; these must be killed by
an autoclave. However, for many of the projects in this book, the
glassware can be sterilized by boiling as described above.

Always remember that a sterile jar can become nonsterile when
exposed to air. If the jar is not hot and is open to the air, it can become contaminated. The longer a container is open, the more likely
it will become contaminated. Whenever you conduct microbiology
experiments, consider where and how the sterile glassware and tools
could get contaminated and do your best to prevent this from hap-
pening, since it will ruin your project.

To sterilize media mix purchased from a supply house, follow the
instructions that accompany the mix. This is usually done with an autoclave, but other techniques are available.

Inoculating loops

Transferring microbes from a vial or dish to a sterile culture media is
called inoculation. You must transfer them on a sterile utensil; an inoculating loop (shown in Fig. 1-4) is usually used for this purpose.
The loop is simply a metal (usually platinum) or plastic loop, connected to a handle. The plastic, presterilized inoculating loops available from scientific supply houses are designed to be used once and
thrown away.

Permanent

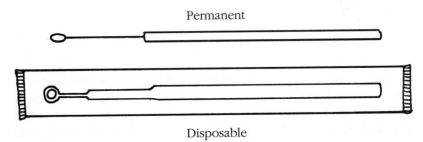

Disposable

1-4 *Long-lasting inoculating loops must be sterilized in a bunsen
burner between each use. A disposable loop is used only once and then
disposed of.*

If you are using a platinum inoculating loop, it must be "flamed"
in a bunsen burner (Fig. 1-5) to become sterilized before each use.
The loop is placed in the flame until it becomes red-hot. Let the loop
cool for 30 seconds or so. (If the loop is too hot when it enters the
culture, it will kill the cells and the transfer will not be successful.) Do

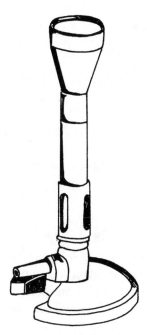

1-5
*The bunsen burner is used
to sterilize inoculating loops
and fix stains on slides.*

not wave the loop around in the air when it is cooling or lay it on the countertop before use. This would recontaminate it.

 Caution! Always work with your teacher or advisor when a project calls for the use of flames.

Once the loop cools, transfer a sample of the culture (called *inoculum*) from the source to the vial or plate containing sterile culture media. Dip the loop into the vial or touch the loop to the surface of the plate containing the culture. Some of the culture will adhere to the loop.

To transfer it to a vial containing broth, simply stick the loop into the sterile broth and shake it around. To transfer it to a plate containing agar, gently move the loop over the surface of the media. Don't dig the loop into the surface of the agar. (See Fig. 1-6.) The same method is used with an agar slant.

Once you have inoculated the culture, discard the loop if it is disposable or reflame it if it is metal. If you put the loop down on the counter after the transfer, you contaminate the countertop and spread microbes around the room. Therefore, always put the loop back into the flame after it has been used as well as before use.

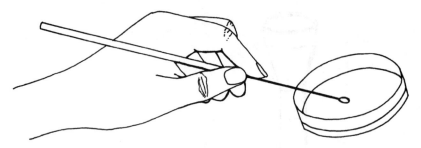

1-6 *Gently touch the inoculating loop to the growth media. Don't dig the loop into the agar.*

Spreading rods

A spreading rod, shown in Fig. 1-7, is required for some of the projects in this book. This is simply a metal or glass rod with a bend in it. The longer, top portion is the handle, while the bottom portion, beneath the bend, is used to spread a liquid culture (inoculum) over the surface of an agar plate.

1-7 *A spreading rod is used to make a consistent layer over the entire culture plate.*

⚠ **Caution!** Sterilizing a spreading rod must be performed by your teacher or advisor since it is a dangerous procedure.

The spreading rod must be sterile or the culture will become contaminated. This is usually done by dipping the spreading rod in alcohol and lighting the rod to burn off any contaminants. This is dangerous if not done correctly, so let your sponsor or other experienced person handle it.

To sterilize a spreading rod, your teacher or advisor will dip the bottom portion in alcohol and ignite the alcohol on the loop with the flame until all the alcohol has burned off. The spreading rod *must* be angled downward while the alcohol is burning so it doesn't drip on the person holding it. Keep the rod away from the container of alcohol while it's burning. Also keep the container of alcohol away from the open flame.

Use the sterile spreading rod to gently spread the culture equally over the surface of the plate. Do not set the spreading rod down on the work area, since it will contaminate the table.

Pipettes

A pipette is a long, thin tube with markings that show the volume of liquid in the tube. (See Fig. 1-8.) A pipette is used to transfer liquids from one vessel to another. When a pipette is used to transfer a liquid culture of microbes, it must be sterile or the resulting culture will be contaminated.

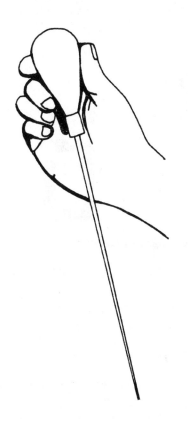

1-8
Never pipette with your mouth. You can use a bulb (as seen here) or a pump.

Sterile, disposable pipettes can be purchased from a scientific supply house. They are not expensive and can save a lot of lost time or failed experiments because of contaminated pipettes. Sterilization of pipettes is difficult.

Caution! You should never pipette by mouth. Pipetting by mouth is dangerous because some of the fluid can get into your mouth. Also, it is too easy to contaminate a culture when pipetting by mouth.

Use a pipette bulb, which works just like an eyedropper bulb, to bring the fluid into the pipette by suction. Some pipettes use a pump instead of a bulb to suck liquids into the tube. It might take a little practice to accurately transfer small volumes of liquid with a pipette bulb or pump, but it is well worth your time to practice the safe use of pipettes.

Some of the experiments in this book require a *micropipette*, which accomplishes the same thing as a regular pipette, but allows you to transfer much smaller quantities of liquid. A micropipette is shown in Fig. 1-9.

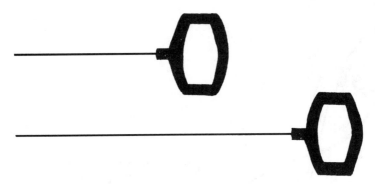

1-9 *Micropipettes are used to draw millionths of a liter.*

Growth media and sterile technique are necessary to see and study the tiny, fascinating world of microbes. However, the most important tool that has aided the study of these microbes is the microscope.

Microscopes

Microscopes magnify objects, letting us see objects smaller than our eyes would permit. Our ability to see is limited by our eyes' *resolving power*, the ability to see two separate objects that are extremely close to one another. If two objects appear to be one, then your eyes have not "resolved" them. Microscopes give much greater resolving power, or resolution, than our eyes alone can.

The most common type of microscope is called the *compound microscope*. One is shown in Fig. 1-10. These microscopes have one

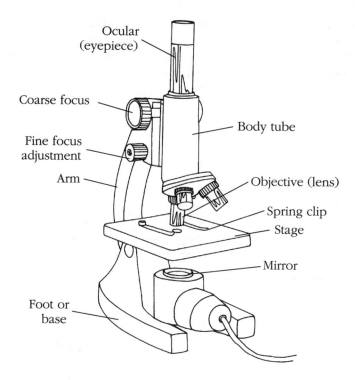

Ocular
(eyepiece)

Coarse focus

Body tube

Fine focus
adjustment

Arm

Objective (lens)

Spring clip

Stage

Mirror

Foot or
base

1-10 *You should be familiar with the basic parts of a micro-scope.*

or two eyepieces, called *oculars*, that you look through to see the magnified image of the object. The ocular contains a lens that usually magnifies the microbe 10 to 12 times. If the scope has only one eyepiece, it is called a *monocular scope*. If the scope has two eyepieces, it is called a *binocular scope*. (Don't confuse a binocular microscope with a stereoscope, also called a dissecting scope, mentioned later in this section.)

In addition to the ocular, a compound microscope contains at least one *objective lens*. The objectives magnify the power of the ocular. Most high school microscopes have a low-power objective (10X), a high-power objective (40X), and an oil-immersion objective (100X). (Oil immersion is discussed in more detail in the next section.) If you have a 10X ocular and a 40X objective, your total magnification is 400X (10 times 40).

Microscopes have a *stage* that holds a microscope slide containing the object to be viewed. A drop of pond water, a culture of bacteria, or a sample of mold growing on bread can be placed on a microscope slide, covered with a coverslip, and then placed on the

microscope stage. The stage has a hole through it, so light can pass
through. The stage can have a mechanical device attached to it, called
a *mechanical stage*, as shown in Fig. 1-11. This stage has knobs and
a fine ruler that allows better control when moving the slide about on
the actual stage.

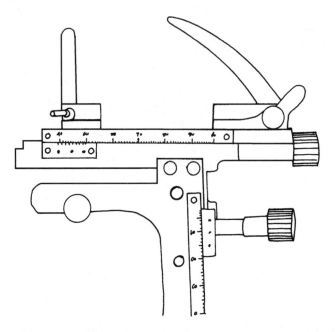

1-11 *A mechanical stage makes it easier to scan the field of vision.*

On the side of the microscope stage are knobs that adjust the distance between the slide and the objective, so you can bring the image into focus. Some scopes have a single coarse focusing knob, but
others have a second, smaller focusing knob for fine adjustments.

Beneath the stage is a *substage*. Here, the light from a light source
is condensed and directed up through the hole in the stage and
through the object being observed. The light source can be sunlight
that is directed through the substage with a reflecting mirror or a
lamp.

Once the light passes up through the object being viewed, it continues up through the objective lens to be magnified, through the ocular lens to be magnified once again, and finally to your eye. (See Fig.
1-12.)

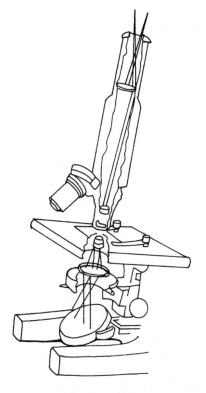

1-12
*To understand how a
typical compound
microscope works, follow
the path of light.*

A few experiments in this book require a *stereoscope*, also called a *dissecting scope*. (See Fig. 1-13.) These scopes have two objectives as well as two oculars, producing a three-dimensional (stereoscopic) view of the object. A stereoscope is a low-power, high-resolution microscope designed to view objects such as entire colonies of microbes growing on a petri dish, pond water, or mold growing on food.

Proper use of a compound microscope

It is always easiest to start viewing through a microscope using the lowest power (10X) objective. It is also easier to focus on the edge of the coverslip, rather than trying to focus on the specimen first. Even when the microscope is completely unfocused, you can still see the edge of the coverslip. Use the large (coarse) and small (fine) focusing knobs to focus on the edge of the coverslip.

Once the edge is in focus, adjust the mirror and the condenser to get just the right amount of light. The amount of light entering through the stage is important; too little light and you won't be able

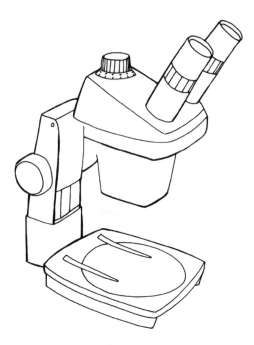

1-13
A dissecting stereoscope is useful for looking at large structures, such as colonies of bacteria.

to distinguish different structures, too much light and everything will be a glare. Most microscopes have a condenser with which you control the amount of light. Once the light is adjusted and the edge of the coverslip is in focus, move the slide on the stage until the specimen is in view. Use the fine adjustment knob (if available) to get the object in perfect focus.

Now that the light is adjusted and the object is in focus, turn the objective wheel to the next-higher power, usually 40X. The object will still be approximately in focus. Use only the fine focusing knob (if available) from this point forward. Notice that the higher-power objective is longer than the lower-power objective, and the oil-immersion objective is the longest of them all. Because there is very little room to move the stage up and down (to focus) with the high-power objectives, there is a danger of coming down too far and cracking the slide (and damaging the objective). Therefore, try to do most of the focusing at the lower power, where there is room for movement. Don't let the objective lens contact the coverslip.

With an oil-immersion objective, light passes up through the stage and microbe, and then through oil instead of the air. This allows you to use a far higher objective—usually 100X, resulting in 1000X magnification (100 times 10). You must use the oil-immersion objective to see bacteria cells. Some of the projects in this book call for the use of an oil-immersion objective.

To switch to the high-power oil-immersion objective, first be sure to have the object in focus under the high-power (40X) objective. Move the objective out of the way and put a drop of immersion oil on the center of the coverslip. Now move the oil-immersion objective into place. Watch this from the side of the stage, to be sure the objective won't be crunched into the slide. There is just a small amount of room between the oil-immersion objective and the slide. This space should now be filled with oil.

Clean off the oil after use with lens paper. Do not get oil on the dry objectives. After using lens paper to take off the majority of the oil, use a tiny drop of xylol on lens paper to wipe off any remaining oil. Do not use alcohol to clean objectives; it can damage an objective.

Incubation

Incubating a culture means keeping the plates or vials that contain microbes at the proper temperature to encourage growth. This is usually done by placing them in an incubator (shown in Fig. 1-14) which maintains a constant temperature. Your school probably has an incubator. If not, it might be possible to run some of the experiments in this book by incubating the cultures in a warm part of a room or under a lamp.

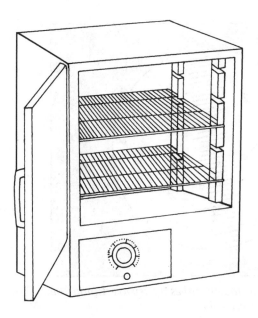

1-14
An incubator is essential in microbiology.

If you are not using an incubator, be sure to test the temperature of the area to be used before beginning any experiment. Do this by placing a thermometer in the area and reading it throughout a day and night. It is important that the temperature of the area does not fluctuate more than two degrees from the required temperature.

Cell counts

Some of the projects in this book require you to count the number of cells in a certain volume of liquid. The easiest and most accurate way to do this is with a *hemocytometer*. A hemocytometer is a special microscope slide that contains a known volume, under which is a very small grid. Put a drop of the culture you need to count in the hemocytometer, cover it with its coverslip, and then count all the cells in the known area to get an estimate of the total number of cells in a known volume. A hemocytometer is, however, expensive and might not be available at your school.

If your microscope has an ocular micrometer (shown in Fig. 1-15), you can still estimate the numbers of cells present, including bacteria. If no measurement device is available, you can estimate the number of some types of cells, such as most algae, fungi, or protists, by placing a small piece of grid paper (use 1-mm-x-1-mm) beneath the slide. You can then count the number of microbes in that area. By knowing the volume of culture placed beneath the coverslip (let's say 0.1 ml

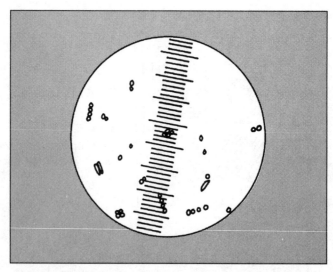

1-15 *An ocular micrometer makes it easy to estimate the size of many types of microbes.*

from a pipette) and knowing the area under the entire coverslip (usually 22-mm-×-22-mm), you can determine a rough estimate of the number of microbes present in the entire 0.1-ml sample.

Zone of inhibition

Many of the projects in this book require you to measure areas on a petri dish that have no cell growth. These areas, called *zones of inhibition,* surround paper discs or pieces of string containing a substance that kills bacteria, such as an antibiotic. The paper discs are placed in agar plates as the bacteria grows. The larger the area around the disc (zone of inhibition) containing no cells, the more antimicrobial is the substance on the disc.

The zones of inhibition can be measured with a ruler, but they can be very small and require a finer instrument such as a *vernier caliper,* shown in Fig. 1-16.

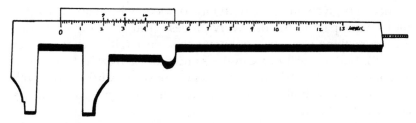

1-16 *A vernier caliper makes it easy to measure the zones of inhibition in many of the experiments.*

2

An introduction to scientific research

Science fairs give you the opportunity to not only learn about a topic, but to participate in the discovery process. You probably won't discover something previously unknown to science (although you never can tell), but you will perform the same process by which such discoveries are made. Advances in science move forward slowly, with each experiment building upon a previous one and preparing researchers for the next.

Advances in medicine, biotechnology, agriculture, and virtually all other scientific disciplines proceed one step at a time. A typical science fair project allows you to see what it is like to take one or two of these steps for yourself.

For example, a question such as whether oregano can control the growth of microbes that spoil canned foods might be answered by a series of experiments. First, lab studies can determine whether oregano has any antibacterial qualities. If it does, an experiment would determine the best quantity of oregano to use per ounce of food. Flavor tests might be run to determine if the oregano changes the flavor of the food to be preserved. Once the preliminary research is complete, field tests can be run using actual canning methods to see if oregano is practical for canned food preservation.

Any one of these experiments might result in bad news. For example, the amount of oregano needed to preserve food might be so high that it interferes with the flavor of the food. Such a study is still valuable, however, since it provides information that keeps scientists on the right track. It might lead scientists to conclude that only foods that would be enhanced by the flavor of oregano could be preserved this way.

Each oregano experiment is necessary before the next can be performed, and the entire progression is necessary before a successful conclusion can be reached. Scientific research, no matter how simple or how sophisticated, must follow a protocol that demands consistency and, most important, duplication. When one scientist or research team finds some new revelation, others must verify it. The scientific method provides a framework for researchers to follow. It assures a highly focused, reproducible sequence of events.

The scientific method

The fundamentals of the scientific method can be divided into five steps. The following paragraphs describe each step and relate them to sections in later chapters of this book.

Step 1: Problem

What question do you want to answer, or what problem would you like to solve? For example, can bacteria become resistant to common household disinfectants, or can shaking someone's hand really result in a transfer of pathogens from a sick person to a healthy person? The "Project overview" section of each project in this book gives a number of questions and problems to think about. The "Going further" and "Suggested research" sections can also give you ideas to spark your imagination about additional problems.

Step 2: Hypothesis

A *hypothesis* is an educated guess, based on your literature search and observation of the natural world, that offers a possible answer to the questions posed. You might hypothesize that bacteria cannot become resistant to disinfectants or that shaking hands can transfer pathogens from person to person. Form a hypothesis about the questions given in the "Project overview" section of each project.

Step 3: Experimentation

The experiment is designed to determine whether the hypothesis is correct or not. If the hypothesis turns out to be incorrect, a well-designed experiment helps determine why.

Any experiment starts with a design process. How must you prepare the experiment and what procedures must be followed to test the hypothesis? What materials will be needed? What live organisms, if any, are needed? What step-by-step procedures must be followed during the experiment? What observations and data must be made and collected while the experiment is running? Once these questions have been answered, the actual experiment can begin.

Next, you must perform the actual experiment, make observations, and collect data. The results must be documented for study and analysis. There are three important things to remember when conducting an experiment: take notes, take notes, and take notes. The more details, the better; the most common mistake new scientists make is thinking they will remember some small detail. If you always carry a notebook and pencil when working on your project, this won't become a problem.

Some science fairs require the project notebook be submitted along with a brief abstract of the project. Some fairs require or encourage a full-length report of the project as well. (See the section on science fair guidelines in chapter 3 of this book.)

The "Materials list" section of each project lists all the materials needed for each experiment, and the "Procedures" section provides step-by-step instructions. Suggestions are given on what observations should be made and what data should be collected.

Another important aspect of experimentation is *replication*. For any project to be considered valid scientific work, the experimental groups should be replicated as many times as possible. Multiple groups can then be averaged together or, better yet, statistically analyzed. For the projects in this book, try to perform all experimental groups in triplicate. Replication reduces the chances of collecting spurious data, which will result in erroneous analysis and conclusions.

Step 4: Analysis

After the experiment is complete, you must analyze the data and draw conclusions to determine if your hypothesis is correct. Create tables, charts, or graphs to help analyze the data. The "Procedures" section of each project suggests what observations to make and what data to collect while running the experiment. The "Analysis" section asks important questions to help you analyze the data and often provides empty tables or charts to fill in with your data. This book provides guidance, but you must draw your own conclusions.

Step 5: Conclusion

The conclusions should be based upon your original hypothesis. Was it correct? Even if it was incorrect, what did you learn from the experiment? What new hypothesis can you create and test? Something is always learned while performing an experiment, even if it's how *not* to perform the next experiment.

Building on past projects

Just as scientists advance the work of other scientists, so too can you advance the work of those who have performed science fair projects before you. Don't copy their work, but thinking of what the next logical step might be in that line of research. You might be able to put a new twist on a previous experiment. For example, if electromagnetic radiation can damage nematodes (small roundworms that live in the soil), does it harm the fungi that the nematodes feed on? Or, if the original experiment was performed *in vitro* (in a test tube), can you perform a similar experiment *in vivo* (in nature)?

Abstracts of previous science fair projects are available from the Science Service in Washington D.C. See appendix B for this and other sources of successful science fair projects. (Also see the next chapter for more help on selecting a project.)

3

Getting started

Since you are looking through this book, we can probably assume you have an interest in microorganisms. Therefore, the first thing to do is find out specifically what piques your interests, if you don't already know. There are a few ways to do this.

The best way to select a project is find out what interests you about microbes, if you don't already know. Did you ever wonder what hidden organisms could be discovered in a handful of dirt or a cupful of pond water? Did you ever stop to think about the microbes carried by that fly on your picnic plate, your friend's hand that you are shaking, or the raw oysters you or your parents love to eat? How about that bottle of "pure mountain water"—just how pure is it?

Use this book

Stop and let your mind wander for a while. What comes to mind? It could be anything, anywhere, anybody! Once you've opened your mind and let your imagination run wild, look through the table of contents in this book for specific topics to research. Select a project that you are not only interested in, but truly enthused about.

This book contains 20 science fair projects about microorganisms. Read through the "Background" and "Project overview" sections of each project. Every project in this book can be adjusted, expanded upon, or fine-tuned in some way to personalize your investigation.

After reading through these sections, think about how you can put your own signature on the experiment. The "Going further" and "Suggested research" sections of each experiment are designed to help you personalize each project. If you find yourself saying, "I'd like to know more" about something you read here, you're well on your way to selecting a science fair project about microbes.

Other sources

At this point, you can begin your project or continue to look for more insight into the problem. Consider branching out by looking through science sections of newspapers. Do you see articles about contaminated drinking water or some new, life-threatening bacteria?

Also, look at magazines such as *Popular Science, Discover,* or *Omni.* Check the *Reader's Guide to Periodical Literature* in your library, which indexes articles in numerous magazines and gives a brief synopsis of each. Your school textbooks might also be helpful. Check references to other books, usually found at the end of each chapter.

Other sources that can help include educational television shows such as "NOVA," "National Geographic" specials, and "Nature." Almost all of these types of shows are found on public television and cable networks. Check your local listings to see what might be showing in the near future in your area. Also, don't hesitate to use past science fair projects as a source of interesting topics.

Talk to specialists in the field

Once you have a good idea for a project, consider talking with a professional. For example, if your project involves using a new kind of chemical-free roach trap to kill cockroaches, speak with a typical homeowner who uses chemical cockroach traps, a professor of entomology who specializes in domestic insects at a nearby university, a representative of the company making the new trap, and a representative from the company that makes conventional traps. Interesting science fair projects don't only involve equipment, chemicals, and cultures, but also what people have to say about the topic: pro, con, and neutral.

Also, be sure to use any special resources that are readily available. For example, if you live near a water treatment plant, hospital, landfill, agricultural research station, farm, or almost other facility that can contribute to your project, use it to your advantage. If you have a parent or friend who is involved in a business or profession applicable to your project, try to incorporate it into your research.

All the projects included in this book are good candidates for science fair projects. To make one of these projects an outstanding example of research, put your own signature on it. Include portions of the "Going further" section and delve into the "Suggested research" section. Perhaps a teacher, scientist, or businessperson can add an interesting aspect to the research that makes it truly unique.

Before you begin

Before starting any experiment, review the entire project with your sponsor to anticipate problems that might arise. Some projects must be done at a certain time of year. Some can be done in a day or two, while others can take weeks, months, or even longer.

Some projects use supplies that are found around the home, but many require equipment or supplies that must be purchased from a local hardware store, science/nature store, or a scientific supply house. Many projects require microbial cultures that might be available from your school or can be ordered from a scientific supply house. Your sponsor might have access to the organisms needed for the project. Organisms such as insects might be caught in the wild, bought at a pet or bait shop, or ordered from a scientific supply house.

Also, plan ahead financially. Look through the "Materials list" section of each experiment. Be sure to add materials for the additions or modifications you are making to the original project. Determine how and where you will get everything and how much it will cost. If a microscope with an oil-immersion objective is required, do you have access to one? If you need a bacterial culture, is it available from your teacher, from a nearby university, or must you purchase it? How much will it cost? Don't begin a project unless you can budget the appropriate amount of time and money as suggested by your sponsor.

Performing the experiment

Once you have selected a project, use the following suggestions to get organized.

Scheduling

Before proceeding, develop a schedule to ensure you have a complete project in time for the fair. Have your sponsor approve your timetable. Leave yourself time to acquire the equipment, supplies and organisms.

Most science fair projects require at least a few months from start to finish if they are to be accomplished thoroughly. It would be difficult to produce a prize-winning project without plenty of time. Here is a chronological list of things to include when preparing a timetable:

- Identify your adult sponsor.
- Choose a general topic and establish a project notebook.
- List resources (libraries to go to; people to speak with; businesses, organizations or agencies to contact).
- Select reading materials and use bibliographies for more resources; begin a formal literature search
- Select the exact project and develop a hypothesis; write a detailed research plan and discuss it with your adult sponsor; have your sponsor sign-off on the final research plan
- Procure equipment, supplies, organisms and all other materials.
- Follow up on your resource list: speak with experts, make all contacts, etc.
- Set up and begin experimentation; collect data and rigorously take notes.
- Begin to plan for your exhibit display.
- Begin writing your report, analyzing data, and drawing conclusions.
- Complete your report and have your sponsor review it.
- Design the exhibit display.
- Write the final report and abstract, and be sure your notebook is available and readable.
- Complete and construct a dry run of the exhibit display.
- Prepare for questions about your project.
- Disassemble and pack your project for transport to the fair.
- When the fair arrives, set up your display, relax, and enjoy yourself!

The literature search

As you can see from the suggested schedule, one of the first things to do is to perform a literature search of the problem you intend to study. A literature search (also often called simply *research*) means reading everything you can get your hands on about the topic: newspapers, magazines, books, abstracts, and anything else related to the specific subject. Use online databases, if available. Talk to as many people as possible who have some insight into the topic. Listen to the news on radio and television. At this point you might want to narrow down or even change the exact problem you want to study.

Once your literature search is complete and you have organized the data both on paper and in your mind, you should know exactly what problem you intend to study and then formulate your hypothesis.

The research plan

At this point you should have completed a research plan. You can use portions of this book to get started with your research plan, but you must go into additional detail and include all modifications. Before beginning the project, go through your plan detail with your adult sponsor to be sure the requirements of the project are safe, attainable, suitable, and practical.

In many science fairs, your sponsor is required to sign off on the research plan attesting to the fact that it has been reviewed and approved. It is important to review your particular fair's regulations and guidelines to be sure your project won't run into any problems as you proceed.

Science fair guidelines

Almost all science fairs have formal guidelines or rules. Check with your sponsor to see what they are. For example, there might be a limit to the amount of money that can be spent on a project or a rule against the use of certain microorganisms. Be sure to review these guidelines and check that the experiment poses no conflicts.

Many science fairs require four basic components for all entries:

1 The actual notebook used throughout the project that contains data collection notes (Be sure to consider this when taking your notes, since they might be read by fair officials.)

2 An abstract of the project—usually no more than 250 words long—that briefly states the problem, proposed hypothesis, generalized procedures, data collection methods, and conclusions

3 A full-length research paper

4 The exhibition display

The research paper

A research paper might be required at your fair, but consider doing one even if it isn't necessary. (You might be able to earn extra credit for the paper in one of your science classes.) The research paper should include seven sections:

1 A title page

2 A table of contents

3 An introduction

4 A thorough "Procedures" section explaining what you did

5 A comprehensive "Discussion" section explaining what went through your mind while performing the research and experimentation

6 A conclusion that summarizes your results

7 A reference and credit section listing your sources and giving credit to any individual, company, organization, or agency that assisted you.

Look in the bibliography for a list of books that detail how to write a research report.

The display

The exhibit display should be as informative as possible. Remember, most people—including the judges—will only spend a short time looking at each presentation. Try to create a display that gets as much information as possible across with the least amount of words. Use graphs, charts, or tables to illustrate data. As the old saying goes, "A picture is worth a thousand words."

Discuss with your sponsor any exhibit requirements such as special equipment, electrical outlets, and wiring needs. Live organisms including microbial cultures of any kind are usually prohibited from being displayed. Often, preserved specimens are also prohibited. Usually no foods, wastes, or even water is allowed in an exhibit. Also, no flames, gases, or harmful chemicals are allowed. Find out what you can and cannot do before proceeding.

Many fairs have specific size requirements for the actual display and its backboard. For more information on building an exhibit display, see the bibliography at the end of this book.

Judging

When beginning your project, remember that adherence to the scientific method and attention to detail is crucial to the success of any project. Judges want to see a well thought-out project performed by a knowledgeable individual who understands all aspects of the project.

Most science fairs assign a point value to various aspects of a project. For example, the research paper might be worth 25 points, while the actual display might be worth 5 points. Request any information that might give you insight about the judgment criteria at your fair. This can help you allocate your time and resources where they are needed most.

Part 2

Controlling & using microbes

This section contains three projects that investigate unique and unusual ways of controlling microbes or using microbes to control other organisms.

Do you know anyone who likes to eat raw oysters? Many people do. Since the oysters are eaten raw, bacteria found on and in the oysters are not destroyed by heat during cooking. Many people pour hot sauce over the oysters to spice them up. The first experiment in this section considers whether this hot sauce kills bacteria on the oysters, decreasing your chance of becoming ill.

The second project investigates how a fungus found in nature controls insects and how this natural method of control might be applied to everyday use in a new, "high-tech" roach trap.

The final project in this section looks to tree bark for new ways to control bacterial growth. The tree barks from common trees of North America are tested for their antibacterial qualities.

4

Oysters, hot sauce, & bacteria

Does hot pepper sauce kill bacteria found on raw oysters?

Background

Microbes are found on just about everything we eat. Meat, poultry, and seafoods harbor large numbers of bacteria and other microbes. These microbes might be on the animals when they are caught or slaughtered, or they might get on the food during processing, packaging, or shipping.

Fortunately, most of these microbes are destroyed when food is cooked. Heat is one of the best methods of killing microbes, including disease-causing bacteria. Many people, however, eat foods such as oysters raw. Oysters are bottom feeders, meaning they eat decaying matter off of the ocean's floor. Therefore, they contain many microbes when they are harvested. Pathogenic organisms that might be found on these oysters have the potential to make someone sick.

Project overview

Some people like to put hot sauce on their oysters. Are these people less likely to ingest bacteria on the oysters than people who don't use hot sauce? The purpose of this experiment is determine whether hot pepper sauce kills bacteria on raw oysters.

Hot pepper sauce contains a substance called *capsaicin*. Capsaicin is the active ingredient that puts the "hot" in hot sauce. Does a hot sauce such as Tabasco sauce (which contains capsaicin) kill bacteria on raw oysters? Will more hot sauce kill more bacteria?

Materials list

- Petri dishes (at least eight)
- Beakers
- Blender
- Autoclave
- Bunsen burner
- Tweezers
- Nutrient agar (from a scientific supply house)
- Seawater (Collect your own or reconstitute dehydrated sea salts, available in a supermarket.)
- Sterile filter paper disks (available from a supply house)
- Tabasco sauce (Use a new bottle; it loses its punch with age.)
- About a dozen raw, fresh oysters in their shells
- Eye dropper
- Adhesive tape
- Fine ruler (or vernier calipers)

Procedures

Sterilize all the glassware including beakers, petri dishes, eye droppers, and the blender jar in an autoclave (or use the sterilization technique described in chapter 1). The nutrient agar and seawater must also be sterilized prior to starting the project.

To prepare the petri dishes, start with liquid nutrient agar. You can create your own from a mix or melt down solidified agar in a beaker. Once the agar is in liquid form, add 1% seawater to the agar. For example, if you are working with 500 ml of liquid agar, add 5 ml of seawater and mix thoroughly.

While the agar is still hot and in liquid form, pour it into each of the eight sterilized petri dishes. Fill each dish only about ⅓ full. Wait for the agar to solidify.

Once the agar is solid, use sterile forceps or tweezers to place a filter paper disc on the agar in the middle of the first petri dish, as shown in Fig. 4-1. Pass the tweezers through a bunsen burner flame a few times before picking up the next blotter paper disc and placing it on the next petri dish. Continue for all the dishes, passing the tweezers through the flame each time to keep them sterile. The Tabasco sauce will be applied to these discs.

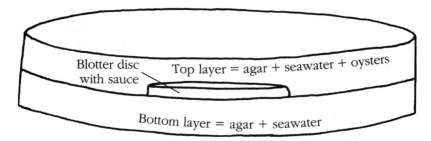

4-1 *The bottom layer contains agar and seawater; the paper and hot sauce is in the middle; the agar, seawater, and blended oysters are on the top layer.*

Use a sterile eye dropper to place one drop of Tabasco sauce on the disc in the middle of a petri dish. Create another identical dish. Label both dishes *1 drop.* Put two drops of the sauce on another disc in a petri dish. Create another identical disc and label them appropriately. Place three drops on discs for two other dishes and label them as well. Leave the last two petri dishes alone without any sauce to act as the controls; label them as such. Let the sauce dry for at least three hours before continuing.

You are now ready to add raw oysters to each petri dish. Purchase fresh, raw oysters and shuck them (or have your sponsor shuck them) as you normally would. Prepare another batch of liquid agar as you did earlier by mixing a small amount of seawater with the agar. You'll need 125 ml of agar mixed with 1.5 ml of seawater. While the agar is being readied, prepare the oysters.

Place about one dozen oysters (include both the juice and the meat) in the blender and add 100 ml of seawater so the juice it produces isn't too thick. Blend thoroughly. Using a sterile graduated pipette, transfer about 10 ml of the oyster juice into the liquid agar that you just prepared and mix thoroughly.

Wait for the oyster/agar blend to cool down slightly, but don't let it solidify. Pour a thin layer of the agar/oyster juice blend onto the existing agar layer (and filter paper) in each of the eight petri dishes. Allow this layer of agar to solidify. Be sure all the petri dishes are labeled with the proper number of drops of Tabasco sauce placed on the filter paper.

Leave all the petri dishes at room temperature for 48 hours. The bacteria present in the oysters will grow throughout the nutrient agar. If the bacteria are destroyed by the sauce, there will be a halo around the paper discs containing the sauce, as shown in Fig. 4-2. This is the zone of inhibition (see chapter 1 for more information). Use a fine ruler or vernier calipers to measure the zone of inhibition around each disc in each plate. Observe all the petri dishes under a stereoscope if available. Record your data.

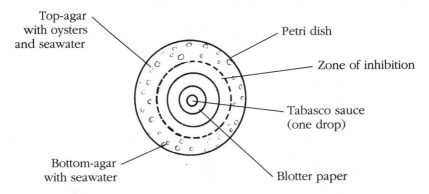

4-2 *Look for the zone of inhibition around the blotter disc that contains the hot sauce.*

Analysis

Did the discs with sauce have zones of inhibition around them? Did more sauce result in larger zones? Does it appear that hot sauce has some antibacterial qualities? Does the use of hot sauce on raw oysters reduce the number of bacteria you eat and therefore reduce the risk of becoming ill when eating raw oysters?

The two control dishes should have bacterial growth all around and right up to the edge of the discs. If they don't, there was not enough bacteria present at the onset of the project and you should re-run the project using more oysters in greater concentration. You could also incubate the dishes to ensure a constant temperature.

Going further

- What would happen if you replaced the seawater with distilled water in this project? Will it work as well, if at all?

- Devise a similar experiment to test for natural bacterial inhibitors in flavor enhancers used on beef or poultry. For example, does A-1 steak sauce have any antibacterial effect on beef?
- Look at the bacteria that grew on the plates in this project under an oil-immersion objective and try to identify the types of bacteria. Does the sauce work on many forms of bacteria, or just one?

Suggested research

- Research the epidemiology of eating raw foods. How many people get sick from raw foods? Is this a serious problem?
- Look into the new process of irradiating foods to keep them from becoming contaminated. What are the advantages and disadvantages of this process?

5

Microbial pest control

Using fungus to control insect pests

Background

Many microbes are parasites that must live on or in another organism (the host) to survive. Some parasites don't harm their hosts, others cause illness and disease, and some even cause the death of the host. Microbial parasites often control the populations of their hosts. For example, an outbreak of the gypsy moth caterpillar was recently controlled in Connecticut by a fungal parasite. The fungus attacked the caterpillars, killing them before they could pupate and turn into adult moths. Since the caterpillars never made it to adult moths, there were few eggs laid and fewer gypsy moths for the next few years until they gradually made a comeback.

For parasites to thrive, however, they must have the correct environmental conditions. In the example of the gypsy moth caterpillars, excessive and long-term humidity in the region lead to the success of the fungal parasite.

Project overview

Microbial parasites such as this fungus are found in nature. Fungi are beginning to be used to our advantage in waging war against insect pests. A similar fungus is being used to control household insect pests.

There are two parts to this experiment. In the first, you determine what environmental conditions are needed for a fungal parasite to infect and kill crickets. In the second, you see if these same conditions affect the efficiency of a new roach trap that uses a fungus instead of toxic chemicals to kill cockroaches.

Materials list
For part 1

- 4 empty cottage cheese containers (16 ounce) with lids
- Scissors
- Newspaper
- A few fresh apples
- 8 crickets (Collect your own or purchase from a pet store or supply house.)
- Balance
- Spatula
- Freeze-dried culture of *Beauveria brassicae* (available from a scientific supply house)
- Water sprayer/mister
- Thermometer
- Cool area in building
- Warm area in building

For part 2

- 4 large cardboard boxes with lids (at least 4 feet square or larger)
- Newspaper
- Packaging tape
- 4 dog biscuits
- A few fresh apples
- 4 Bengal Roach Chambers (These are available from stores that sell roach traps. You might have to ask the store to special order these traps from EcoScience in Worcester, Massachusetts; they cost about $6 each.)
- About 40 cockroaches (Catch them on your own or purchase them from a supply house.)

Procedures
Part 1

Use the scissors to punch four small airholes in the cottage cheese container lids. Cut the newspaper so it can lay flat on the bottom of the container. Place a layer of newspaper on the bottom of each container. Put one small apple slice (less than $\frac{1}{16}$ of apple) in each container and add two crickets to each container.

Inoculate the containers by measuring on the balance one gram of the freeze-dried *Beauveria brassicae* with the spatula. Place one gram of this fungus into each container. Put the lids on all the containers and label them as follows: *cool/dry, cool/moist, warm/dry,* and *warm/moist.*

Place the two *cool* containers in a cool area of the building (such as the basement) and put the two *warm* containers in a warm area or under a lamp. Be sure the temperature under the lamp doesn't go above 28°C. Every day, use the water sprayer to very lightly mist the newspaper on the floor of the two containers labeled *moist.* Every two to three days, replace the apple slice with a fresh slice.

Each day, check the containers. (If you can't find the crickets, check under the newspaper.) If a cricket dies, note the changes in its appearance. Leave the dead cricket in its container and continue daily observations. Note the time of any external changes to the cricket's body. Does it appear "moldy"?

Part 2

The second part of this project involves performing a similar experiment using cockroaches as the subjects and unique roach traps as the source of the fungus. Bengal Roach Traps (shown in Fig. 5-1) contain fungal spores that are picked up by insects as they pass through the trap. The trap contains a pheromone that attracts the insects.

Open up the cardboard box lids. Seal all the edges of the box (except for the top) with packaging tape so the cockroaches cannot escape through the cracks. Line the bottom of each box with newspapers. Place one roach trap in each of the four boxes. Position the traps in the same location in each box. Add a dog biscuit to each box for food at the opposite end from the trap, and place four thin apple slices in each box. The apples will provide additional food and moisture. Write on the outside of each box one of the following labels: *cool/dry, cool/moist, warm/dry,* and *warm/moist.* (See Fig. 5-2.)

Use the mister to gently spray the newspaper in both of the boxes labeled *moist.* Apply the same number of sprays to both boxes. Fi-

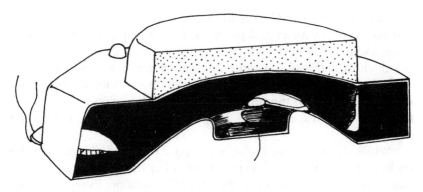

5-1 *This new type of roach trap uses pheromones to attract cockroaches and fungi to kill them.*

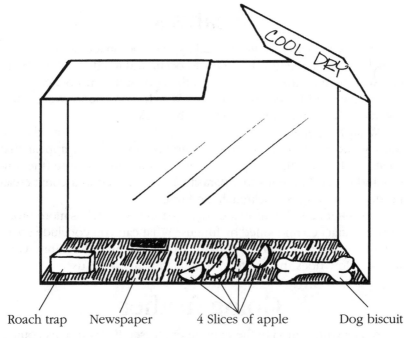

COOL DRY

Roach trap   Newspaper   4 Slices of apple   Dog biscuit

5-2 *The cardboard box is lined on the bottom with newspapers and contains the roach trap, a dog biscuit, and four slices of apple.*

nally, add 10 cockroaches to each of the four boxes and immediately close the lids. Use packaging tape to seal closed the tops. Poke 10 small holes in each side of each box for circulation. Be sure the holes are not large enough for the insects to crawl out.

Place the two *cool* boxes in the same cool location used in the first part of this project and place the two *warm* boxes in the same warm location. Every day, carefully open the seal of the two boxes labeled *moist* and add the same amount of water mist to each box. Bang the top of the boxes before opening them to knock down any roaches crawling on the tops of the boxes. Also open the other two boxes for the same amount of time, even though you are not adding moisture.

While the boxes are open, make your observations. Look at the cockroaches carefully, looking for any kind of growth. Count how many insects have grown and their overall condition. Every third day, replace the apple slices in all four boxes. Use a magnifying glass or stereoscope to look at the cockroaches that die. Do you see any growth?

Analysis

In part 1, were the crickets attacked by microbes in all the containers? How long did it take for the attack to occur and to affect all the crickets? Was there a difference between the cool and warm boxes or between the moist and dry boxes? Which conditions were most favorable to the microbes' attack on the crickets?

What happened in the second part of the project? How did the traps work in the four different environments? Does it appear that what you learned about a parasite and host found in nature (the fungus and cricket) pertains to a parasite and host used in a commercial product (the fungus, cockroach, and trap)?

Do certain conditions make crickets and cockroaches more prone to being attacked and killed by fungus? What can you conclude about the effect of temperature and humidity on the growth of fungus on insects? How does this relate to fungus that grows on us?

Going further

- For part 1, modify the experiment to see if this fungus could be used not only in the home, but also in the field. For example, what would happen if it were used in a small-scale garden under controlled conditions? Would it attack beneficial insects as well as pest insects? How about using other types of insects?
- Both subjects in this project were adult forms. Would the fungus attack any larval forms, such as caterpillars?

- For part 2, run an experiment to compare the success of the Bengal trap, which uses a natural fungus to control pests, and more conventional traps that use toxic chemicals.

Suggested research

- Research how this trap works and compare the health and safety aspects of using this trap versus more conventional traps. Contact EcoScience for more information about how it attracts cockroaches and how it kills the pests.
- Investigate the use of parasites and parasitoids to control other insect pests.

6

Trees & bacteria

Can tree bark inhibit bacteria growth?

Background

For thousands of years, many cultures around the world have used trees and shrubs for medicinal purposes. Today, many of our medicines come from plants. Recently, the bark of the Pacific Yew tree was found to contain a substance used to treat some forms of cancer. When a plant species becomes extinct before it can be studied, a cure for some disease might be lost forever.

If extracts from plants can inhibit the growth of nonpathogenic bacteria, continued research might determine that the substance can also control pathogens.

Project overview

This project investigates whether the bark of four North American trees can inhibit bacteria growth. The trees included in this study are Engelmann Spruce (*Picea engelmannii*), Manzanita (*Arctostaphylos*), Ponderosa Pine (*Pinus Ponderosa*), and White Fir (*Abies Concolor*). All of these species are abundant in North America. Some of these trees grow to over a hundred feet and live for 250 to 500 years.

If these trees are unavailable in your area, you can still perform this experiment by selecting other species of trees and following the same procedures given here.

This experiment is designed to see if the bark of these trees can inhibit growth in *E. coli* and *Bacillus subtilis*. The project is divided into two parts. In the first part, you prepare filter paper discs (similar to antibiotic discs) containing extracts from each type of tree bark. In the second part, you prepare the microbial cultures and apply the discs to test whether the extracts can control the bacteria.

Materials list

For part 1

- Small samples of such tree bark (enough to produce about 1 gram of sawdust) as the following:
  ~Engelmann Spruce
  ~Manzanita
  ~Ponderosa Pine
  ~White Fir
- Knife or peeler
- Blender
- Balance
- Four 40-ml beakers
- Distilled water
- Graduated cylinder
- Parafilm
- Paper hole-puncher
- Filter paper
- 250-ml beaker
- Tweezers or forceps
- Aluminum foil
- Marker
- Autoclave (optional)

For part 2

- Culture of *E. coli* on an agar slant (available from a scientific supply house)
- Culture of *Bacillus subtilis* on an agar slant (available from a scientific supply house)
- Latex gloves
- 24 blood agar plates (also available from a supply house)
- Inoculating loop
- Bunsen burner
- Hydrogen peroxide

- Incubator
- Metric ruler
- Vernier calipers (optional)

Procedures
Part 1

Remove any leaves from the bark samples. Use a peeler or knife to strip away the outer layer of bark until the bark is off-white in color and has a smooth texture. Cut the first sample into very small pieces (so the blender is not damaged) and place it in the blender. Grind the bark into a fine sawdust. Collect the sawdust in a beaker. Use a balance to measure 0.5 grams of the sawdust and place it in a 40-ml beaker.

Use a graduated cylinder to measure 7 ml of distilled water and add it to the beaker containing the bark. Label the beaker with a marker and cover it with parafilm. (See Fig. 6-1.) Thoroughly clean the blender and dry it so none of the sample remains.

6-1 *Each beaker contains a different bark extract.*

Repeat this procedure for each of the bark samples. You'll end up with four 40-ml beakers. These beakers will be the source of the extracts used to create the filter paper discs.

Use the paper hole-puncher to punch out 150 discs out of the filter paper. Place all of these discs into a 250-ml beaker filled with 100 ml of distilled water (Fig. 6-2). Cover this beaker with aluminum foil and place it into an autoclave at 121°C for 15 minutes to sterilize.

Remove the parafilm from each of the four beakers and use sterile tweezers or forceps to place 35 of these discs into each of the four bark solution beakers. (A different pair of tweezers must be used for each bark, or the tweezers must be sterilized between beakers.) Cover all four beakers with aluminum foil and place them in an au-

6-2
The paper-punched discs are placed in distilled water before being soaked in the four bark extracts (stored in small beakers).

toclave at 121°C for 15 minutes to sterilize. Remove the beakers from the autoclave and set them aside for now.

Part 2

Using sterile technique as described in chapter 1, inoculate four blood agar plates with *E. coli*. Have your sponsor show you how to streak the plates using a consistent radiating method as seen in Fig. 6-3. Label each plate, indicating the tree used. One disc (from the same bark) is placed on each radius, as shown. Repeat this procedure with the other three plates for each of the other tree barks.

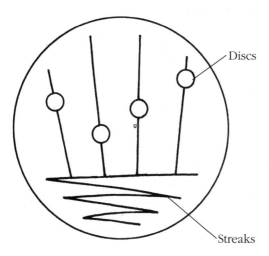

6-3
The agar is streaked as shown, and the paper-punched discs are placed on each of the radiating streaks.

 Caution! To maintain sterility, have your sponsor dip the tweezers into a 3% hydrogen peroxide solution and then flame it before each disc is placed onto the plate. This procedure is dangerous.

Repeat the entire procedure for four more plates inoculated with *B. Subtilis*. Use the same streaking method to inoculate the plates and the same technique to prepare and apply the discs. Once you have eight plates ready—two sets (from the two types of bacteria) of four plates (for each bark tested)—you can incubate the plates. Place all eight plates in an incubator for 24 hours at 37°C.

Create two more identical sets of these eight plates. Incubate the first set for 36 hours and the second set for 48 hours. At the end of each time period, use a fine ruler or vernier calipers to measure the zone of inhibition on each plate. The zone of inhibition looks like a halo surrounding the discs where colonies of microbes do not grow.

Analysis

Which bark extracts produced the largest zone of inhibition around which of the two bacteria? Did the most inhibition occur after 24, 36, or 48 hours? Do any of the extracts appear to have some inhibitory effect on one or both of the bacteria? If so, did this effect increase or decrease over time?

Going further

- If one of the extracts could inhibit growth, find related species of that tree and test them as well.
- If the effect appeared strongest with the least incubation time, repeat the experiment with even shorter incubation periods to see if it increases the effect even more.

Suggested research

- Investigate what chemicals are present in plants (or specifically bark) that inhibit bacterial growth.
- Read about the number and importance of medicines originally derived from botanicals (plants).
- Read about old herbal remedies used for thousands of years and alternative medicines used today. Do folk doctors know something that Western science is just learning about today?

Part 3

Microbes on the move

This part contains three projects that investigate how microbes get around. The first project looks into a rather unusual method of transport—one that most people wouldn't think about. Do insects such as flies and ants carry microbes on their feet and deposit these microbes when they walk over your picnic plate?

The second project focuses on how you can accidentally spread microbes from one person to another. Can you spread microbes to many people by simply shaking hands? Are microbes really spread by sneezing or coughing? This project investigates both of these questions.

The final project looks at the most common method of microbe transport: the wind. Are different numbers and types of microbes carried on the wind at different times of day and during different seasons?

7

Insects & microbes

Can insects transport microbes?

Background

The term *germs* is often used to describe microbes such as bacteria, viruses, and fungi that cause disease. As mentioned in chapter 1, these so-called germs are properly called pathogens. When trying to control pathogens, it is important to understand how they are transported from place to place. Many pathogens can simply float through the air, landing on unsuspecting organisms. When you sneeze or cough, you launch microbes, possibly including pathogens, into the air so they can travel a long distance.

Microbes are also spread by mechanical means. For example, placing contaminated ground beef on a countertop in your kitchen spreads the pathogens to the countertop, which can then spread pathogens to other foods that come in contact with the same countertop. When you eat this contaminated food, you ingest microbes that came from a cow hundreds or thousands of miles away.

Project overview

Microbes hitch rides on the wind or in the water. They can also hitch a ride on other organisms. How often have you seen a fly or other insect land on your food during a picnic or on your kitchen counter during the hot summer months? (See Fig. 7-1.) Have you ever been in a restaurant and seen a cockroach crawl across the floor or up a wall? Where do you think it was walking before you saw it? Can organisms, such as insects, transport microbes and possibly pathogens from one place to another? Do different insects carry different numbers or types of microbes?

7-1 *Do you think an insect is depositing microbes as it crawls across your picnic plate?*

Materials list

- About a dozen sterile nutrient agar plates (available from a scientific supply house)
- A variety of insects caught in their natural habitat (you can improvise this list):
  ~Ants
  ~Cockroaches
  ~Houseflies
  ~Crickets
  ~Any other insects you might find in your home or on your food during a picnic
- Small screw-cap glass jars to hold insects
- Forceps
- Adhesive tape
- Marker to write on plastic

- Microscope
- Autoclave (optional)
- Camera (optional)

Procedures

Prepare one sterile glass jar for each insect to be tested. Use the autoclave or the technique described in chapter 1 to sterilize the glassware.

Capture at least one of each insect to be tested and place it in a sterilized jar. Try to catch insects such as flies, ants, or cockroaches in your home or in a restaurant. Each insect should be placed in a sterile jar and remain there for as short a time as possible before proceeding with the next step. All the insects should remain in their respective jars for the same amount of time. Label each jar with the name of the insect to be tested.

To begin the actual experiment, prepare the insect to be transferred to a nutrient agar plate. For crawling insects such as ants, this is easy. Simply let the insect crawl out of the jar onto the plate. Close the agar plate with the lid and let the ant walk around on the agar for one minute. Then, pick the ant up off the plate with the sterile forceps and release or dispose of it.

Repeat this procedure with the other insects in different agar plates, each appropriately labeled with the type of insect. For flying insects, shake the jar containing the fly just before dumping it out onto the plate. This will stun the fly. If it gets stuck in the agar, use a sterile glass rod to free it. Once on the plate, the fly will primarily walk around on the agar since there won't be enough room for it to fly. (See Fig. 7-2.)

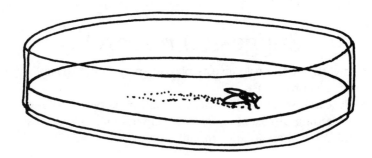

7-2 *The fly will crawl over the agar since there won't be enough room for it to fly about.*

After each insect walks over its respective plate, seal the plate with adhesive tape and incubate it at 30°C for 24 to 48 hours. After that time, observe the plates for microbial growth. Take photographs (if a camera is available) of each plate for comparison. Use a magnifying glass or a stereoscope to see the colonies and to count the number of colonies in each plate. The colonies will appear along the path where the insects walked on the agar.

Count all the colonies along the trail. Then divide the trail into one-inch-long segments. Count the number of colonies in each one-inch segment. Take the average of all the segments and use this number to compare with the other insects.

Analysis

Did all the insects track microbes across the plates? Which insect was responsible for producing the largest number of colonies per inch? Did the appearance of the colonies differ among the different insects? Did different types of microbes grow in different plates?

Going further

- Look at the structure of the insect's "feet" (called *tarsi*) under a stereoscope or with a magnifying lens. What might contribute to their ability to carry microbes?
- Continue the project to see if the insect's original habitat changes the numbers of microbes transmitted to the plates. For example, collect some flies around a dump, in the woods, and in your home and compare the microbial growth produced on agar plates. Compare these insects with an insect reared in a contained environment such as an aquarium with a prepackaged food supply.

Suggested research

- Investigate the role insects play in spreading disease. Is it a serious problem?
- Research whether the insect's mouthparts play a role in spreading disease. Is a fly's mouthparts more likely to spread microbes than an ant's mouthparts?

8

Spreading germs

Does shaking hands or coughing really spread bacteria?

Background

When many people in a community are sick, you will often hear people say, "It's going around," meaning the "germs" (pathogens) are being spread throughout the community. How are pathogens transmitted from sick person to person?

Bacterial cells are invisible to the naked eye, so you can never really tell where they are and when they are being transmitted from one person to another. Two methods of transmitting bacteria cells from one person to another are by physical contact and through the air, since they are small enough to be carried on wind currents.

Project overview

This project is divided into two parts: studying microbe transmission by contact and studying them through the air. The first part investigates whether microbes are transmitted from one person to another by a simple handshake. If so, how many people can be "contaminated" from a single source? (This project uses harmless yeast cells instead of actual pathogens.)

In the second part of this project, you'll see how far air can transmit microbes when a person coughs (or sneezes).

Materials list
For part 1

- Package of baker's or brewer's yeast (Fleischmann's yeast is usually available in supermarkets)
- Sugar
- Pond or tap water
- Large beaker (about 600 ml)
- 15 petri dishes
- 15 Sabouraud agar plates (available from a scientific supply house)
- Box of disposable surgical gloves (from a supply house)
- Box of sterile cotton swabs
- Eyedropper
- Incubator
- At least five individuals (including yourself)
- Marker

For part 2

- 11 sterile nutrient agar plates (from a supply house)
- Meter or yardstick
- Bench or table that is at least waist-high and 6 feet long
- Another, smaller table or bench
- Two people to assist you
- Surgical face masks (from a medical supply store, scientific supply house, or doctor)
- Adhesive tape

Procedures
Part 1

First, prepare an active culture of yeast cells by adding 50 grams of sugar to 450 ml of tap or pond water in a large beaker. Then add the contents of one bag of Fleischmann's yeast (or a teaspoon of baker's or brewer's yeast) to the beaker. Stir the liquid and leave the beaker in a warm part of the lab or your home for 24 to 48 hours. Do not cover the beaker. After this time, your beaker should contain an active culture, loaded with yeast cells.

Divide the petri dishes into three sets of five. Label the three sets *#1* through *#5*. Set aside two of the sets.

Have each of the five people put a surgical glove on his or her right hand. Each person should be assigned a number, with you being #1. Use an eyedropper to place a few drops of the yeast culture on your glove. Rub the liquid all around your glove with a sterile swab. This contaminates your glove with the yeast. You are now the source of a simulated pathogen.

Next, shake hands with the #2 person. The #2 person then shakes hands with the #3 person. This is repeated until the fourth person has shaken hands with fifth. As soon as all the handshaking is completed, have each person touch the fingertips of their gloves to the proper petri dish from the first set, as shown in Fig. 8-1. The fourth person, for example, touches the #4 petri dish. Place the entire set of dishes in an incubator for 24 hours at 37°C. (If an incubator is not available, place the dishes in a warm place for a similar period of time.)

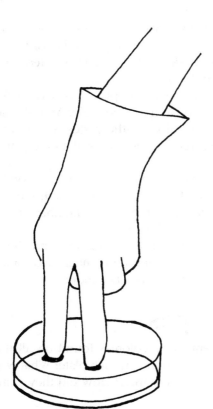

8-1
Touch your gloved fingers to the agar.

Repeat this entire procedure two more times using new gloves and the other sets of petri dishes. After 24 hours have passed, observe the three sets of five dishes for culture growth. Count the number of colonies on each plate and record the data. The colonies should look like white, shiny, circular mounds on the plate. Mark the plate's lid above each colony as it is counted; this will prevent you from counting the same colony twice.

Average the number of colonies for each set of plates. For example, average the number of colonies found in all three of the #5 plates. Then compare the averages of the different plates.

Part 2

For this part of the project, you need a room with very little air movement. Label 10 of the agar plates as follows: ½, 1, 1½, 2, 2½, 3, 4, 5, 6, and *control.* Set up the 6-foot table and place the smaller table a few feet behind it, as shown in Fig. 8-2. Place the appropriately marked, closed agar plates at each of the distances listed (½ foot, 1 foot, etc.). Place the 10th (control) plate on the table behind the long one as shown.

Stand at the end of the 6-foot table (between the two tables, as in Fig. 8-2), facing it. Have two people stand along the edge of the 6-foot table wearing surgical face masks so they aren't infected with your "germs" and don't inoculate the plates.

Have your helpers quickly remove all the plate lids (except for the control plate) and step back away from the table. You should then immediately cough in the direction of the plates. Have your helpers wait 30 seconds and then replace the lids. Seal all the plates with adhesive tape along their edges.

Wait a few minutes, then open the control plate for 30 seconds, but don't cough into it. There are always some microbes in the air. The control will give you an estimate of the normal number of microbes in the air.

Incubate all the plates in an incubator or a warm area and check at 24 hours and 48 hours for growth. Count the number of colonies on each plate. Repeat the experiment at least once more.

Analysis

In the first part of the experiment, were colonies found on all the plates? If so, how many colonies were found on each plate? What was the average for each plate from the three trials? How did they com-

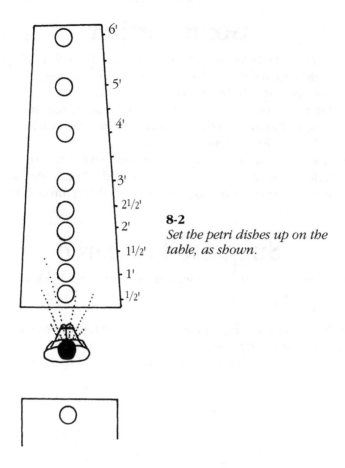

8-2
Set the petri dishes up on the table, as shown.

pare with each other? Did the numbers of colonies remain constant, decrease, or increase from the lowest- to the highest-numbered plate?

Does it appear that physical contact, such as shaking hands, can transmit microbes from one person to another—even to people who never had direct contact with the original source of the microbes? If yeasts can be transmitted, is it possible that infectious diseases can be transmitted?

For part 2 of the experiment, what do you conclude about the distance microbes can be carried by a cough? Subtract the number of colonies found in the control from all the experimental plates. Did the number of colonies diminish with increased distance from the source (you)? Is there a safe distance to be from someone sneezing or coughing?

Going further

- What other ways are pathogens spread? Look through your bathroom and try to devise an experiment to see if microbes are being spread from one family member to another. Establish your own disease-prevention guidelines for your class or family, and then compare them with established public-health guidelines.
- Repeat part 2 of this experiment, but use a throat spray before you cough, or cover your mouth. Does a throat spray really reduce the number of microbes flying through the air after a sneeze?

Suggested research

- Research the history of infectious disease epidemics such as the Bubonic plague of long ago, and other diseases like AIDS today.
- Speak with your family doctor or other health professional about how simple sanitary techniques can reduce the spread of disease. Relate them to this project.

9

Microbes in the wind

Do the numbers & types of microbes carried by wind change over time?

Background

Air is more than just a collection of gasses. It carries many particles, some of which are living. Large particles fall out of the air, but smaller particles can be carried for hundreds of miles.

Since microbes are so small, the wind can play an important role in their dispersal. Spores from molds that grow on land have been found hundreds of miles out to sea. Bacterial cells from a sewage treatment plant can be found in the air over 100 feet away from the plant. Microbial spores have even been found at altitudes of 90,000 feet.

Project overview

Weather plays a role in determining how many and what kinds of particles are found in the air. Would you find different numbers and types of microbes in the air at different times of the day because of changes in wind, temperature or humidity?

How might microbes in the wind differ during the four seasons? Seasonal changes in most living creatures are easy to see. Many trees lose their leaves in the fall, grass gets greener in the spring, flowers bloom in the summer, but does the community of microbes found in the air change with the seasons?

This project is divided into two parts. In the first, you will see if the numbers and kinds of microbes found in the air differ at different times of the day. In the second part, you'll look for changes that occur due to changes in season.

Materials list

For part 1

- 24 sterile nutrient agar plates (available from a scientific supply house)
- Marker
- 1-inch adhesive tape
- Timer
- Microscope (with an oil-immersion objective)

For part 2

- 36 sterile nutrient agar plates
- 36 sterile Sabouraud agar plates (available from a scientific supply house)
- Incubator
- Drawing paper

Procedures

Both parts of this experiment should be done on days with little or no wind and no precipitation.

Part 1

Label one set of nutrient agar plates as follows: *6 a.m., 10 a.m., 2 p.m.* and *6 p.m.* as shown in Fig. 9-1. Make six sets, totalling 24 plates. At 6 a.m., go outside and hold open the appropriate plate for exactly one minute. Do not move the plate around while it is open.

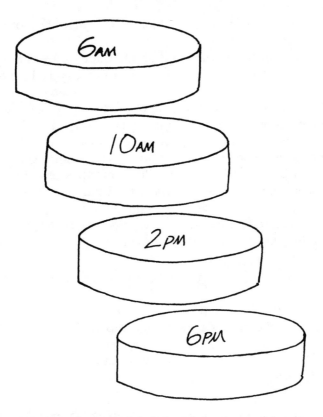

9-1 *The plates are labeled with the times of day at which they will be opened.*

Close the plate and seal it with adhesive tape around the edges. Put the plate in an incubator at 32°C for 24 hours.

Repeat this procedure at 10 a.m., 2 p.m., and 6 p.m. Observe the sealed plates 24 and 48 hours after they were exposed to the air. (Since they were opened at different times, you will be making these observations at different times.) Note the number of the colonies and their form.

Repeat this procedure on five different days (there should be no wind or rain on any of the days). After looking at the gross structure of the colonies, use the oil-immersion objective to study the types of bacteria present.

Part 2

This part will take about three months to perform. Follow the procedure described in the following paragraphs once each week for a three-month period. It must be done at the same time of the day throughout this period.

Take three plates of nutrient agar and three plates of Sabouraud agar into your backyard. Open one plate at a time and pass it through the air by waving your arm back and forth twice, then close the plate. Do this for all six plates and seal all of them with adhesive tape. Label the bottom of the plates with the date and type of media (nutrient or Sabouraud). Incubate the plates for two days at 37°C.

After the incubation period, note the shapes and colors of the colonies on each plate. Draw the different types of colonies. You need a detailed, written description of their forms, since you will be comparing colony types over a three-month period. After observing the gross structure of the colonies, use the oil-immersion objective to study the individual types of cells present.

Analysis

For the first part of this project, compare the growth on the plates at different times of day. Plot a graph similar to the one in Fig. 9-2. Does it appear that there are different amounts and types of microbes floating in the air at different times of day? What might be the reason for these differences?

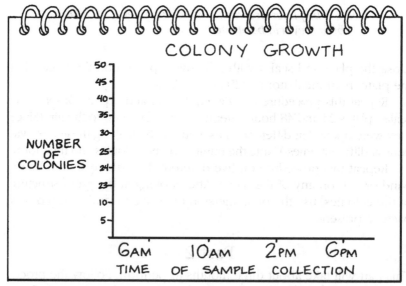

9-2 *Use your data to fill in a chart similar to this one.*

For the second part of this project, look for changes in the types of colonies over the three-month period on one type of media. Did the numbers and kinds of microbes differ throughout this period? Plot a graph of the results for each type of media.

Then, compare the differences between the two types of media. Do you see the same types of colonies on both of the plates incubated in the same week? Are there differences between the two groups over the period? Do the different media give rise to different types of microbes? If so, all or just some of the time? What do you conclude about seasonal change in air-borne microbes?

Going further

- Continue part 1 of this experiment by taking samples throughout the night. Do you get any different types of growth? Repeat both parts of this project with different types of culture medium. Use media specifically designed to culture different types of microbes. Don't just look for bacteria.
- Repeat the second part during different seasons. For example, if you ran the project from summer to fall, try it again from winter to spring.
- Change the elevation of the samples. For example, repeat the first part of the project, but take all your samples at ground level. Or, contact some friends who live in another part of the country and have them run the same experiment and then compare notes.

Suggested research

- Research allergies and their causes. What airborne microbes are responsible for different types of allergies, both outdoors and in the home?
- Investigate whether microbes are associated with air pollution in any way.

Part 4

Garlic, health, & microbes

Many people believe that garlic has medicinal qualities. They believe it has some effect in treating arthritis, other arteriosclerotic conditions, asthma, cancer, colds, digestive problems, flu, heart disorders, high blood pressure, sinusitis, ulcers, yeast infections, and other infections. Quite a bit of research confirms many of these beliefs. Doctors have used garlic since World War I to prevent gangrene and the spread of infection.

The next three projects all look into the ability of garlic (and a few spices) to control the growth of bacteria. The first project uses the vapors from garlic to control bacterial growth. The vapors from a few spices are also tested. Which works the best?

The second project tests the antibacterial effects of garlic juice. The final project in this section goes a step further: Does garlic lose its medicinal qualities if it is heated first? If you cook garlic before eating it, are you getting any of its health benefits?

10

Common spices & microbes

Can vapors from common spices control microbial growth?

Background

When foods are grown or raised, processed, packaged, distributed, and stored on the store shelf, they are candidates for contamination. A barrage of antimicrobial chemicals is used throughout this entire chain of events. Fungicides are used on crops and during packaging to prevent molds, synthetic antibiotics are injected in animals to prevent infections, and fruits are waxed to extend shelf-life by preventing rotting. Almost all packaged foods contain preservatives to extend shelf-life.

The public is becoming increasingly aware of the health and environmental problems caused by many of these chemicals. This awareness began with our understanding of the dangers of insecticides such as DDT and continues today with the recent banning of the fumigant methyl bromide, which was used for years to prevent molds from growing in packaged foods.

Because of this increased public awareness, scientists are searching for alternative ways to protect our foods. One possibility is the use of natural food products to prevent microbes from attacking our food supply.

Project overview

The answer to most of the problems with food additives is not to simply stop using fumigants, fungicides, and preservatives. This would only result in a dramatic loss in our food supplies. The answer is to find substances that will still protect our foods, but pose less of a threat to our health and to the environment.

The purpose of this study is to determine if vapors from common spices can reduce or eliminate bacterial growth. In this project, cultures of *Bacillus subtilis* are exposed to a variety of different spice vapors. By comparing the resulting cultures with a control, you can see if the spices controlled the bacterial growth.

Materials list

- Nutrient broth mix (available from a scientific supply house)
- 2 test tubes
- Nutrient broth culture of *Bacillus subtilis* (available from a supply house)
- 2 bottles of agar
- 10 petri dishes
- 6 different spices, finely ground (create your own list, or use the following: garlic, cloves, cinnamon, horseradish, mustard, onions)
- 30 sterile cotton swabs (available from a supply house)
- 1 vial of disinfectant
- Stereoscope (optional)

Procedures

Prepare two test tubes with sterile nutrient broth. (See chapter 1 for details on sterilization.) Use a sterile pipette to inoculate these two test tubes with one drop of the *Bacillus subtilis* culture. Incubate the two new vials at 37°C for 24 hours. Refrigerate them after the incubation period.

Again using the sterile technique, pour the sterile nutrient agar into seven sterile petri dishes to a depth of 0.5 cm. (If you must liquefy the agar, place the jar containing the solid agar in a pan of water so that the water level is above the agar level. Boil the water until the agar dissolves. Cool the agar slightly before pouring by placing it in a 46°C water bath for a few minutes.)

After the agar solidifies, turn the plates upside down. Label each of the plates with the name of the spice. To place a spice into each plate, raise the bottom half of the petri dish (which is now the top, containing the agar) and place 1 gram of the spice on the lid (which is now the bottom). See Fig. 10-1. Do this quickly to minimize contaminating the plates with the air.

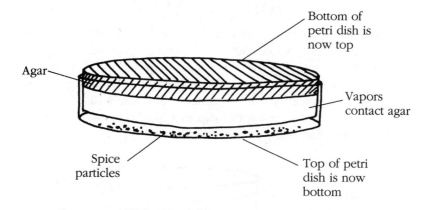

10-1 *The petri dish is inverted so the agar is on the top and the spices are on the bottom. The vapors from the spices will fill the chamber.*

You will now inoculate each of these plates with the bacteria. Dip a sterile cotton swab into one of the two bacteria broth cultures you created earlier and inoculate the first plate by gently swabbing the surface of the agar. Once again, keep the plates open for only a brief moment to prevent contamination from the air. Repeat this procedure for each plate. When all the plates contain the spices and have been inoculated with the bacteria, tape them shut.

Incubate the plates at 37°C for 24 hours and then make observations over the following three days. Look for the number of colonies growing (upside down) in each plate. Count the number of colonies in each plate over the course of the three days and record all the data. Use a stereoscope, if available, to see the colonies in more detail. Calculate the final results.

This experiment should be repeated two more times. The data from all three runs should then be analyzed.

Analysis

The control plates should contain a substantial amount of growth since these plates had no vapors to inhibit growth. How did the plates with the spices do? Did any show a significant reduction in growth compared to the control? Did the inhibition occur immediately after incubation, after three days, or somewhere in between? Do the vapors appear to take a long time to work, or do they begin to lose their effect after a period of time?

Going further

The most obvious way to continue this project would be to test more spices on different types of bacteria. But you can also take this project a step forward by testing it in a small-scale application. For example, devise a new experiment that would apply your results from this project to a field test with jarred or canned foods. Use an apparatus like the one in Fig. 10-2 to see if the vapors with the most promise from this first project can really preserve foods longer.

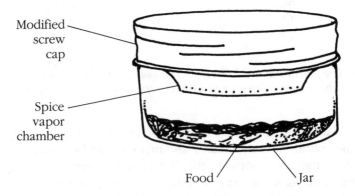

Modified screw cap

Spice vapor chamber

Food

Jar

10-2 *Devise an experiment that will test the vapors in a jar containing food.*

Suggested research

- Research the use of natural foods, such as spices, as preservatives. Are any currently in use in the food-packing industry?
- Look into the use of natural vapors as a pesticide.

11

Garlic juice & microbes

Does garlic juice have antimicrobial qualities?

Background

Garlic, or *Allium sativum*, has been noted as a medicinal herb for centuries. Archaeologists have found ancient Hebrew, Greek, Roman, Babylonian, and Egyptian texts referring to garlic as a cure for many illnesses. Some of these texts date back as far as 500 B.C.

Garlic is a complicated substance made of many different compounds. Research shows that garlic contains unsaturated aldehydes, allicin, allyl disulfides, calcium, copper, germanium, iron, magnesium, manganese, volatile oils, phosphorous, phytocides, potassium, selenium, sulfur, vitamin A, vitamin B1, vitamin B2, and zinc.

Garlic can lower blood pressure because one of its components, menthyl allyl trisulfide, dilates the blood vessel walls, allowing blood to flow more freely. Garlic can also lower serum cholesterol levels and inhibit blood coagulation, thus lowering the risk of heart attacks and blood clots. It's also considered a good aid for digestion.

The allicin in garlic has an antibiotic action that is equivalent to one percent that of penicillin. Garlic is also an antifungal agent that is effective against candidiasis, athlete's foot, vaginal yeast infections, and many other fungi.

Project overview

Garlic appears harmful to the reproduction of bacteria. Allicin is believed to be the active ingredient, even though its mode of action is not known.

E. coli is found normally in the human intestinal tract, where it usually produces no ill effects. When this bacteria leaves its natural home of the large intestine and migrates to other parts of the body, however, it can cause peritonitis, endocarditis, and urinary tract infections. It can also cause infectious diseases, including those associated with the respiratory tract and gastrointestinal tract.

E. coli can be accidentally spread from the human bowel to other parts of the body during periods of diarrheal disease. A breakdown in routine hygiene, such as not washing your hands, allows the bacteria to be transmitted in a fecal-oral manner. The bacteria can then move through the bloodstream and cause other infections.

Can garlic help protect us from the occasional spread of *E. coli* throughout our bodies? The purpose of this experiment is to find out if garlic juice has a natural inhibitory effect on *E. coli* growth.

Materials list

- Garlic juice or fresh garlic
- Bottle of distilled water
- Five 50-ml beakers
- 35 sterile blood agar plates
- Sterile applicator sticks
- Culture of *E. coli* (available from a scientific supply house)
- Swabs
- Large test tube of saline solution
- Bottle of 0.5 McFarland standard solution (available from a hospital supply store or a scientific supply house)
- Saline solution (same as above)
- Scissors
- Small ball of thin string
- Forceps
- Incubator
- Metric ruler
- Camera (optional)

Procedures

The first step of this experiment is to prepare the solutions of garlic juice that will be used on blood agar plates. Sterilize all the glassware as described in chapter 1 before beginning this project.

You can either purchase garlic juice or create your own. To prepare the garlic extract, place 80 grams of fresh garlic in a blender containing 200 ml of sterile water and blend for 20 seconds. You will prepare five different, 20-ml samples of garlic solutions in small beakers labeled: 0% garlic, 25% garlic, 50% garlic, 75% garlic and 100% garlic. (See Fig. 11-1.)

11-1 *You'll have five small beakers, each containing a different concentration. Place eight pieces of string in each beaker.*

For the 0% solution, use 20 ml of distilled water with no garlic. This will be the control. For the 25% garlic solution, mix 5 ml of garlic extract with 15 ml of distilled water. For the 50% solution, mix 10 ml of garlic extract with 10 ml of distilled water. For the 75% solution, mix 15 ml of the garlic extract with 5 ml of distilled water. And finally for the 100% garlic solution, use 20 ml of garlic extract.

Now prepare a test tube containing a known number of *E. coli* cells. This known amount of cells will be used as the standard to inoculate the blood agar plates. You will use a 0.5 McFarland standard as your guide in creating this concentration of cells. (This method is commonly used by medical personnel to estimate cell counts.)

Touch a sterile applicator stick to the surface of the plate containing *E. coli* bacteria and drop the applicator into a test tube containing standard hospital saline solution. Stir the tube thoroughly and compare the *turbidity* (cloudiness) to the 0.5 McFarland standard. If the two solutions don't look the same, continue to add colonies of *E. coli* until they are similar. When the two are the same, the test solution will have approximately 100,000 *E. coli* cells per ml of saline. This tube will be used to inoculate the blood agar plates.

To inoculate the plates, saturate a sterile cotton swab in the tube of bacteria you just created. Gently wipe the swab over the surface of a blood agar plate. Be sure there is a uniform density of bacteria on the plate. Repeat this procedure for all 35 plates.

Now that the plates have been inoculated, prepare the string, which will be soaked in the extracts and placed on the blood agar plates. Cut 140 pieces of string, each 15 mm long. Saturate 35 pieces of string in each of the five beakers containing garlic juice extracts, allowing the pieces of string to soak in these beakers for 30 minutes.

Use sterile forceps to place four pieces of string on each plate as shown in Fig. 11-2. There will be seven plates for each concentration of garlic. As each plate is completed, label the lid appropriately. After all the plates are prepared and labeled, note the appearance of the plates, or take pictures of them. Then place all the plates in an incubator for 24 hours at 37°C.

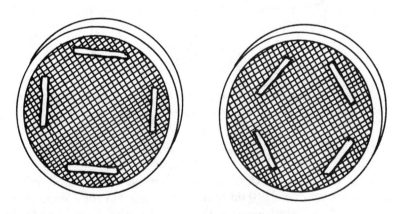

11-2 *Each blood agar plate will have four pieces of string.*

After the incubation period, remove the plates from the incubator. Use a ruler or vernier calipers to measure the width of the zones of inhibition for each plate. This is the area around each piece of string that appears somewhat like a halo and contains no bacterial colonies. After all the measurements have been recorded, photograph the plates again if you are using a camera.

Analysis

Record all the data and calculate the average size of each plate's zone of inhibition. Fill in a table similar to the one in Fig. 11-3. Did you find a difference between the experimental groups and the control group? Did higher concentrations of garlic result in larger zones of inhibition?

Does it appear that garlic might be useful in treating infections caused by *E. coli* bacteria?

GARLIC CONCENTRATIONS	0%	25%	50%	75%	100%	
ZONE OF INHIBITION IN:	MILLIMETERS					
TRIAL #						
1						
2						
3						
4						
5						
6						
7						
8						
9						
10						
11						
12						
13						
14						
15						
16						
17						
18						
19						
20						
21						
22						
23						
24						
25						

11-3 *Record your data in a table similar to this one.*

Going further

Continue this experiment, but incubate the bacteria for 48 hours and check the plates at intervals throughout the duration of the experiment. Is the garlic more effective over time, or does it lose its antibiotic qualities over time? How would this affect its medicinal use?

Suggested research

- Research how doctors and alternative healthcare professionals use garlic as a medicine.
- Look into ways in which people use garlic for its health benefits.
- Research the biochemistry of garlic's medicinal qualities.

12

Hot & cold garlic

Does heating garlic reduce its ability to control bacteria?

Background

The ancient Greeks, Romans, Egyptians, and Chinese all recognized garlic (Fig. 12-1) for its medicinal properties. For centuries, it has been prescribed as a cure for gastrointestinal problems, asthma, "madness," tumors, and parasitic worms.

Recent studies have been conducted to determine if garlic really possesses its legendary medicinal properties. Studies done in Italy and China on populations who routinely ate garlic showed that the incidence of stomach cancer was 60% less than in those whose diets did not include garlic. A similar study in India showed that garlic consumers suffered 45% fewer deaths from heart attacks than those who did not eat it. Studies have shown garlic also contains antifungal properties and inhibits the growth of *Staphylococcus*, *Streptococcus*, *Vibrio*, and *Bacillus* bacteria.

Project overview

Garlic contains a substance called *alliin*, which is found in the solid garlic bulb. This compound, however, has no antibacterial properties of its own. When the garlic bulb is crushed, alliin is converted into *al-*

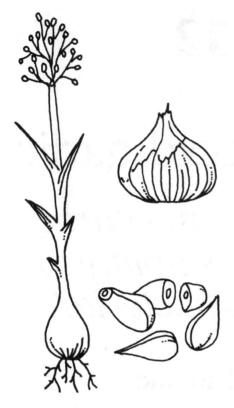

12-1
Garlic (Allium sativam*) is
believed to have many
medicinal qualities.*

licin, which is the active ingredient believed to be responsible for
garlic's medicinal qualities.

This investigation compares bacterial growth inhibition produced
by heated and unheated extracts of garlic. Boiled extracts of garlic and
extracts kept at 4°C are tested. Does the heating process somehow
harm garlic's medicinal qualities? The project continues to determine
whether the inhibitory qualities differ among varying concentrations
of garlic.

Materials list

- Fresh garlic
- Blender
- 12 test tubes
- Large and small graduated cylinders
- Sterile water
- Pot to boil water
- Container to hold an ice bath

- Ice
- Nutrient agar or mix (available from a scientific supply house)
- Stock culture of *E. coli* (available from a scientific supply house)
- 8 petri dishes
- 2 test tube racks
- Graduated micropipettes
- Vernier calipers
- Lab homogenizer (optional)
- Centrifuge (optional)
- Autoclave (optional)

Procedures

To prepare the garlic extract, place 80 grams of fresh garlic in a blender containing 200 ml of sterile water and blend for 20 seconds, creating a juice. Use a graduated cylinder to measure 20 ml of this juice and pour it into a test tube. Place this test tube in a boiling-water bath until it begins to boil. This will be the stock extract for the heated garlic.

Use the graduated cylinder to measure another 20 ml of the garlic juice into another test tube. Place this test tube into an ice bath. Maintain the ice bath at 4°C by adding ice as needed. This will be the stock extract for the cool garlic.

The next step ensures that only the liquid portion of the juice is used for the test, since pieces of garlic in the heated extract might not have been thoroughly boiled. (This step can be skipped, but it might affect the results.)

Place the heated extract into a lab homogenizer for 30 seconds to completely liquefy the juice. Then place the juice into a centrifuge for 10 minutes at 10,000 rpm. This separates any remaining small particles left after the homogenization. The test tube now contains a small amount of sediment at the bottom and a liquid (called the *supernatant*) above the sediment.

Pour off the supernatant and place into a sterile test tube. Repeat this process for the cool extract. Label the proper tube *heated* and the other *cool*. Maintain these two test tubes at 4°C in the ice bath you created earlier.

Now the only difference between the two extracts is that one was boiled prior to the preparation while the other was not. Save both extracts for later use.

Next, prepare the nutrient agar mix according to the instructions that accompanied the mix. Once the mix is created, sterilize it in an autoclave at 15 pounds-per-square-inch pressure at 121°C. (There are alternative ways to sterilize the mix without an autoclave. See chapter 1 for details.) Pour the agar into petri dishes to a depth of about 5 mm and let it solidify.

Pour 3 ml of the remaining agar into each test tube. Make one test tube for each petri dish. Sterilize all the test tubes at the same temperature and pressure as the nutrient agar mix. Let the test tubes cool to room temperature but don't allow the agar to solidify.

Once the test tubes are at room temperature, use a sterile inoculating loop to inoculate each test tube from the stock culture of *E. coli* and mix thoroughly. You now have a mixture of agar and bacteria in each test tube. Pour this mixture out of each test tube, over the solidified agar.in each petri dish. This assures an even distribution of bacteria over the plates.

You will now use the heated and cooled extracts to create extract discs. Use a sterile micropipette to draw 5 microliters of the heated extract. Release the contents onto a sterile filter paper disc. Hold the filter paper disc with sterile forceps and place it on one of the inoculated plates. Repeat this procedure with three other plates. Next, prepare discs using 10 microliters of the same extract and place these discs in the same four plates, as shown in Fig. 12-2. Finally, repeat the process once again, but use 20 microliters of the extract.

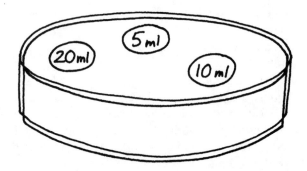

12-2 *Each petri dish has three discs containing different volumes of extract.*

Label the four plates you just created *heated* and label each disc in the plate. You'll have four plates marked *heated* with three discs in each. These different quantities will help you determine if the amount of extract is important, as well as comparing the heated versus cool extracts.

Next, repeat the entire procedure with the cool extract, creating four plates with three discs each. Place all eight plates in an incubator at 37°C for 24 hours.

After the incubation period is complete, measure the zones of inhibition found around each of the discs using vernier calipers as shown in Fig. 12-3. The zone of inhibition is the diameter of the halo that appears around the discs. This halo is the area where bacteria does not grow. Record your results for analysis.

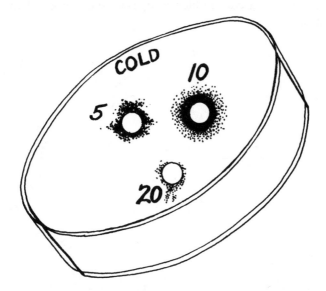

12-3 *Measure the zones of inhibition around each disc in each plate.*

Analysis

Did either the heated or cool preparations inhibit the growth of the bacteria. How did they compare? What would it mean if the cool extract inhibited the bacteria but the heated extract did not? If either did have an inhibitory effect, did the different quantities of extract make a difference? Does the degree of inhibition increase as the volume of extract increases, or does it reach a point of diminishing returns?

Going further

If you found that temperature plays a role in garlic's antimicrobial abilities, continue the project to determine the relationship between heat and effectiveness. Try to correlate the temperature (or length of boiling) with the percent of reduced antimicrobial activity in garlic. What other factors might come into play in effecting garlic's medicinal qualities?

Suggested research

- Research the active ingredient in garlic and how it is created.
- Find the most current literature about the medicinal value of garlic.

Part 5

Microbes: they're everywhere

Even though you can't see them, microbes can be found just about everywhere on our planet. We usually only realize they are present when they make us sick. Luckily, most of the hoards of microbes that we live with everyday don't cause illness, and most are important components in their ecosystems.

The projects in this section research the existence of microbes in a variety of habitats, both in nature and in the home. Did you ever drink bottled water? Do you think there are fewer microbes living in bottled water than in regular tap water? The first project determines if there is a difference between domestic bottled waters and water that comes from halfway around the world, or between carbonated and noncarbonated bottled water?

The next project looks at molds growing in your house. Where can they be found in abundance? The third project studies microbes that live on you. Is your body covered with microbes? Do different parts of your body harbor different types of microbes? Does bathing or showering really help remove these microbes? The final project in this section investigates the microbes living in soil. Do different kinds of soils contain different numbers and types of microbes?

13

Bottled water & microbes

A comparison of the purity of bottled waters

Background

Outbreaks of water contamination in major cities across the country have resulted in an increased awareness of the importance of pure, clean water. Some people have responded by boiling their water; others have installed water purifier systems to filter out contaminants. Another popular way that many people are coping with this problem is to drink bottled waters. Many of these bottled waters are imported from Canada, France, Italy, and Poland, among other places. The cost per ounce of these products is extremely high, considering it's water.

Not only are these bottled waters shipped thousands of miles for consumption in many cases, but they can remain on the shelves for long periods of time. Two years of shelf-life is often considered acceptable. Does water shipped halfway around the globe, to remain on a store shelf for a year, have microbial advantages over tap water or local brands of bottled waters?

Project overview

Many bottled waters are advertised as being pure. In this project you will culture any microbes that might be present in a variety of bottled water. You will compare domestic versus imported water, carbonated versus noncarbonated bottled water, as well as tap water. Sterilized water will act as a control.

Is bottled water as clean and pure as it is advertised to be? Are bottled waters bacteria-free? Are bottled waters purer than tap water? Does carbonated or noncarbonated water offer better protection? You can put your own twist on this project by modifying the types of water you test.

Materials list

- Sterilized water (obtained from a pharmacy)
- Tap water that you routinely drink
- Several different brands of bottled water: a few each of domestic and imported, a brand that comes in both carbonated and noncarbonated varieties, flavored, no-salt, etc. (Be sure the seals are not broken on any of the bottles. Try to purchase similar-sized bottles with the same expiration date, if they have one.)
- Sterile cotton swabs (one for each plate)
- Blood agar plates (two for each water sample, obtained from a scientific supply house)
- Incubator
- Inoculating loop
- Bunsen burner
- Distilled water for gram staining (obtained in a supermarket or pharmacy)
- Gram stain kit (contains crystal violet, iodine solution, alcohol, and safranin—available from a scientific supply house)
- Microscope (with oil-immersion objective)
- Microscope slides
- Camera (optional)

Procedures

The first step is to inoculate the blood agar plates. Break the seal on the first bottle of water. Do not touch the rim of the bottle. Pour a small amount of water over a sterile cotton swab and stroke the cotton swab across the agar plate three times using the sterile technique described in chapter 1. Do not stick the cotton swab into the water container or you will contaminate the bottle.

After inoculating the first plate, repeat the procedure with the same bottle on a second plate. (Each bottle will be used to create two plates.) After the two plates are completed, label them and put them into the incubator set to 37°C. Repeat this process with each of the bottled waters to be tested.

In addition to the bottled water samples, create two plates for the tap water and another two for sterilized water, which is the control. Be sure to use a different sterile cotton swab for each water sample. When all the plates are in the incubator, incubate for 48 hours at 37°C. After this time, remove the plates and photograph the colonies growing on each plate or take detailed notes about their appearance. (See Fig. 13-1.)

Count the gross number of colonies you observe on each plate and the approximate size of each. Once you have made a rough estimate of the number of colonies growing on each plate, analyze the kinds of bacteria present by *gram staining*. (See chapter 22 for detailed instructions on gram staining.)

 Caution! Your teacher or sponsor must be present when you perform gram staining.

Start by sterilizing an inoculating loop by holding it in the flame of the bunsen burner. (Alternatively, you can use sterilized, disposal inoculating loops if you prefer.) Use the sterile inoculating loop to scrape a small amount of the bacteria off of a colony on the first plate and gently smear it onto a microscope slide. Sterilize the inoculating loop after each smear is made. Only one colony of bacteria should be smeared on each slide. Do this with each different type of colony found on each plate. Give all the slides that originate from the same plate the same label. For example, you might have four slides labeled *Mountain Clear Water—noncarbonated.*

After all the smears are prepared, they must be stained to determine whether they are gram-positive or gram-negative. First, place a drop of crystal violet stain on the slide. Let the stain sit on the slide for one minute. Then rinse the slide gently with distilled water. Next, put a drop of the iodine solution on the slide and let it sit for 30 seconds. Rinse the slide gently again. Then add a drop of the alcohol solution, which decolorizes the iodine. After 30 seconds, gently rinse off. Finally, the smear is counterstained with a drop of safranin. Rinse this off after 45 seconds.

The slides can now be analyzed. Look under the microscope to determine whether each sample is purple, meaning gram-positive (it retained the crystal violet stain) or pink, meaning gram-negative (it retained the safranin stain).

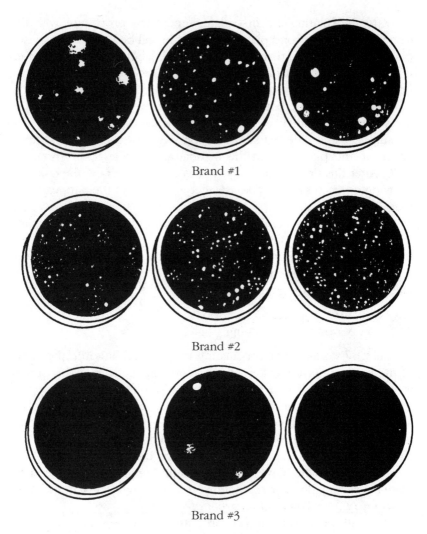

Brand #1

Brand #2

Brand #3

13-1 *Compare the purity of the various brands of bottled water.*

Once you have established the number of colonies and the gram stain, you can determine the shape of each bacteria present. There are three basic shapes: *cocci* (spherical), *bacilli* (rod-shaped), and *spirilla* (spiral-shaped). Look at each slide under a microscope using the oil-immersion objective. Enter your data into a table similar to the one in Fig. 13-2.

BRAND OF WATER	GRAM STAIN	BASIC SHAPE	SPECIES
BRAND 1 NON-CARBONATED	∓	ROD COCCI	ENTERO-BACTER
BRAND 2 NON-CARBONATED	∓	ROD COCCI	
BRAND 3 CARBONATED	+ −	COCCI	
BRAND 4 CARBONATED	+	COCCI	VARIOUS COCCI
TAP WATER CITY WATER SUPPLY TREATED WITH CHLORINE AND FLUORIDE	+	ROD	BACILLUS
DISTILLED WATER	+	ROD	

13-2 *Record your data in a table similar to this one.*

Analysis

The sterilized water samples which acted as your controls should contain no cultures. If they did, your sterile technique did not work properly and the tests must be rerun. If the sterilized water is free from growth, analyze the other samples.

Did the gross count of colonies differ between the tap water and the various types of bottled water? Did the types of bacteria present differ among the samples, both in the gram test and in their shape?

Look for differences not only among the tap water, sterilized water, and bottled waters, but also among the different types of bottled waters. For example, is there a difference between imported versus domestic, or carbonated versus noncarbonated? What are the advantages of buying water from halfway around the world?

Going further

- Run a similar experiment, but test for bacterial growth between bottled waters with no flavoring and those that contain some natural flavoring such as lime, lemon, or raspberry. Does fruit flavoring increase the bacterial count?

- Some brands of bottled waters have more salts than others. Modify the experiment to see if there is a relation between the amount of salts present and the number of bacteria present.

Suggested research

- Research the types of bacteria found in these samples. Are there documented cases of pathogenic microbes being found in bottled waters?
- Look into other contaminants that might be found in bottled waters, especially carbonated waters. How might other byproducts affect the growth of microbes in these waters?
- Research the bottled-water business.

14

Molds in the home

A survey of mold populations in the air conditioner, vacuum cleaner, & home

Background

Molds are a multicellular type of fungi. The main body of a mold is made up of filaments that branch out into the *substrate* they are living on. The substrate might, for example, be a rotting log that provides their food. The body of the mold is called the *mycelium*. Molds reproduce with spores that usually develop on top of stalks called hyphae that rise above the mycelium, as shown in Fig. 14-1.

Mold spores can be carried by air or water. They are commonly found in the soil. If spores travel on the wind or in the water and come to rest in an environment with enough moisture and nutrients, they might grow to form a new mycelium. Spores can be released by the millions. Many humans are allergic to these spores and some spores cause respiratory irritation and even disease.

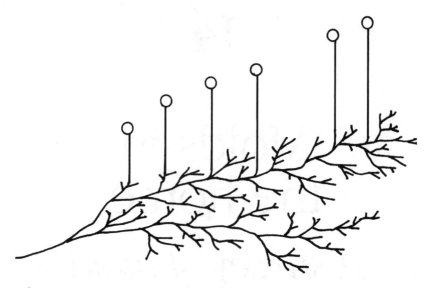

14-1 *Arising from the body of the mold (mycelium) are stalks contain-ing the reproductive spores.*

Project overview

You can assume that mold spores are in your home, but where ex-actly might they be found? Are these spores evenly distributed throughout your house, or are they found in certain parts of a house? Are there places that contain excessively large concentrations of these spores?

This project is divided into three parts. In the first, you do a sur-vey of all the rooms in your home, looking for mold spores through-out. The next two parts look in places that might contain exceptionally high numbers of these spores: an air conditioner filter and a vacuum cleaner bag. Are spores found in these machines? Does the mere existence of spores necessarily mean they will grow? This project tries to answer these questions.

Materials list
For part 1

- 2 Sabouraud media agar plates for each room in your home (available from a scientific supply house)
- 1-inch adhesive tape
- Stereoscope

- Incubator
- Thermometer

For part 2

- Used air conditioner filter (take a used filter out of an air conditioner)
- New air conditioner filter (purchase a new, unused filter for the same air conditioner)
- Scissors
- Marker
- 2 sterile Sabouraud agar plates (available from a scientific supply house)
- 2 sterile nutrient agar plates (available from a supply house)
- Forceps

For part 3

- 12 sealable plastic bags (ziplock type)
- Vacuum cleaner bag (a used, filled bag)
- Teaspoon
- Newspaper
- Clean cloth rag, cut into 12 pieces (each piece 2-x-2-inches big)
- Dry food (dry cereal)
- Moist food (cereal with milk)

Procedures
Part 1

Label the bottom of two Sabouraud plates *kitchen*. Open one plate in your kitchen and wave it back and forth through the air twice. Close the plate and seal it with adhesive tape. Do the same procedure for the other kitchen plate.

Repeat this procedure for each room in your house. Be sure to also do it in the basement, if you have one. Label each set of two plates with the proper room.

Place all the plates in an incubator at 37°C for 48 hours. Twice each day, observe the size of the colonies and the number of colonies on each plate. Draw pictures of these colonies, to show their growth patterns. Use a stereoscope to see the colonies in more detail. Record the data for future reference.

Part 2

Label the bottom of one Sabouraud plate *S used* and the other *S new*. Label the bottom of a sterile nutrient agar plate *N used* and another *N new*. While holding the used air conditioner filter with sterilized forceps (see chapter 1 for information about how to sterilize utensils), cut two 1-x-1-inch pieces from a part of the filter that looks dirty.

Place one of these filter pieces in the center of a nutrient agar plate and the other in the center of a Sabouraud agar plate. (See Fig. 14-2.) Close the plates and seal with adhesive tape. Repeat this procedure with a similar piece of new air conditioner filter. Incubate all the plates at 37°C for 24 to 48 hours.

DIRTY A/C

14-2
Place the piece of dirty air conditioning filter on the agar.

Part 3

Label three of the small plastic bags *dry*, three of them *damp*, three of them *dry food* and the last three *damp food*. Put a few sheets of newspaper down on the floor. Lay the used vacuum cleaner bag on the newspaper. Cut the bag with the scissors so you can insert the teaspoon into the bag. Place 1 teaspoon of vacuum-cleaner-bag contents into each of the labeled baggies.

Soak three of the pieces of cloth in water. Ring them out so they aren't dripping wet, and put one into each of the bags labeled *damp*. Put one of the three pieces of dry cloth into each bag labeled *dry*.

Next, rub 1 teaspoon of dry cereal into another three pieces of cloth (you'll get some dry crumbs of cereal on the cloth) and place each of these into the bags labeled *dry food*. Finally, rub the last three pieces of cloth into the moist food. Put these in the *damp food* bag. Seal all the bags and shake them well.

Incubate the bags in a warm area (about 30°C) of the house for 48 hours. Then look at each sealed bag with a magnifying glass or stereoscope and make note of any growth. Continue the experiment for another two days and repeat your observations. Continue looking at the bags for two more days.

 Caution! Speak with your advisor about how to dispose of these bags.

Analysis

In part 1, look for mold growth in the plates. Did the number of colonies differ among rooms, or were they basically the same? Did any room have more growth than others? What can you conclude about mold distribution in your house?

In part 2, look at the plates after three or four days. Is there any growth? Compare the numbers of colonies. Draw pictures of each plate. Are there differences between the new and old filters? What can you conclude about the ability of the air conditioner filter to collect microbes? What does the air conditioner filter environment provide to the spores that might allow them to grow?

In part 3, did you find mold growth in any of the bags over the two-week period? If so, which bags had growth, and when did it develop? What can you conclude about the presence of mold spores in the vacuum cleaner? Is the vacuum cleaner environment conducive to the growth of spores? What can you conclude about the conditions needed for spores to germinate?

Going further

- For part 1, take samples from your home, but do it during different seasons. For example, do it during the winter when the house has been sealed up to keep the heat in, and again in the summer when the air conditioner has been circulating air throughout the home. Is the distribution of mold affected by house air currents?
- Perform part 1 of the experiment again in an energy-efficient building with very little outside airflow.

- For part 2, repeat this experiment with an air conditioner that has been shut off for more than one month and more than three months. Can spores survive for a long time in an air conditioner filter, waiting to reproduce?
- For part 3, does your vacuum cleaner catch all the spores it sucks in? Hold a damp rag over the vacuum exhaust for two minutes while the vacuum is running as shown in Fig. 14-3, and culture its contents.

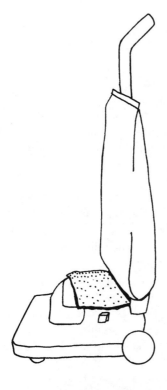

14-3
Place a damp cloth over the exhaust of the vacuum cleaner to catch spores in the exhaust.

Suggested research

- Study the lifecycle of molds and how fungicides kill molds and other fungi.
- Fungi are saprophytes. Study their importance in an ecosystem.
- Study fungi that attack human tissue.

15

Microbes on your body

Where are the most microbes found & what is the best way to control them?

Background

We carry microbes with us every day. Our skin armors us against microbes; it doesn't kill them, but it does prevent them from entering our bodies. Whenever we touch things, we pick up microbes, which explains why hospital staff constantly wash their hands; they don't want to pass pathogens from one patient to another. This also explains why restaurant employees should always clean their hands—so they don't pass microbes from themselves to the food, and then to you.

Project overview

Are microbes present everywhere on your body, or do different areas of your skin contain different numbers (or kinds) of microbes? Are there many different kinds of microbes living on your hands? Does washing your hands and bathing really affect these microbes?

This project is divided into two parts. In the first part, you investigate the cleanliness (or lack of cleanliness) of your hands and the effectiveness of washing your hands with different kinds of soaps. In the second, you go beyond your hands to study the microbes living on other parts of your body and the effectiveness of bathing and showering.

Materials list

- 18 sterile nutrient agar petri dishes (available from scientific supply houses)
- Hot and cold tap water
- Paper towels
- Bar of bath soap
- Bar of bactericidal soap (available at a pharmacy or drugstore)
- 1-inch adhesive tape
- Incubator
- Thermometer
- Stereoscope
- At least 10 sterile cotton swabs (available from a medical supply house, pharmacy, or drugstore)
- Bathtub (or shower)
- A few recently washed bath towels

Procedures
Part 1

Don't bath or wash your hands for at least 4 hours. Open the cover of a sterile nutrient agar plate slightly with your right hand and touch the fingertips of your left hand to the surface of the plate for 1 second. Close the plate and label the bottom *no wash, left*. Repeat this procedure with your right hand and label the bottom *no wash, right*. (See Fig. 15-1.)

Next, rinse your hands under cool running water for 10 seconds and dry them with a towel. Once again, touch your fingers to two new plates as you did earlier. Close these plates and label the appropriate one *rinse, left* and the other *rinse, right*. Then, use a bar of bath soap and lukewarm water to clean your hands for 30 seconds and re-

NO WASH - LEFT

NO WASH - RIGHT

RINSE - LEFT

RINSE - RIGHT

SOAP - LEFT

SOAP - RIGHT

BACT. SOAP - LEFT

BACT. SOAP - RIGHT

15-1 *Prepare eight petri dishes with the labels shown.*

peat the process once again, using another towel to dry your hands. Label these plates *soap, left* and *soap, right.* Finally, wash your hands again, with hot water and bactericidal soap for 2 minutes. Repeat the process one final time for both hands and label the plates *bact. soap, left* and *bact. soap, right.*

Seal the plates by taping their edges closed with adhesive tape. Place all the plates in an incubator at 37°C for three days. Watch the plates each day and make notes about microbial growth on the plates. Each day, note the shape and number of the colonies, their color, position on the plate, and size. Observe the colonies in the closed plates first with the naked eye and then under a stereoscope.

Part 2

Take a sterile swab and rub it over 1 inch of skin on the back of your hand. Immediately roll the swab over the entire surface of a sterile nutrient agar plate, and close plate. Label the plate *b/b—back of hand.* (The *b/b* is for *before bath*.) Repeat, taking sterile swabs from various parts of your body: forehead, underarm, back of knee, sole of foot, etc. (See Fig. 15-2.) Be sure you only take a 1-inch swab from each body section. Label each plate (on the bottom) with the name of body section where the swab was taken, preceded by *b/b*. Now take a bath (or shower) the way you normally do, using the same soap as you usually do.

15-2 *Rub a sterile cotton swab across your forehead.*

After you have dried off, repeat the process by taking swabs in exactly the same regions of your body as you did before the bath or shower. Label these plates *a/b* (for *after bath*) followed by the body region. Now close all plates by taping their edges with adhesive tape and place them in an incubator at 37°C for 48 hours.

Watch the plates each day and make notes about microbial growth on the plates. Each day, note the shape and number of the colonies, their color, position on the plate, and size. Observe the colonies in the closed plates under a stereoscope.

Analysis

For part 1, how did the colonies differ between each group of fingerprints? Was there a noticeable difference in numbers of colonies?

How about different kinds of colonies? Does washing your hands make a significant difference in controlling the numbers and kinds of microbes on your hands? Does using warm water and certain soaps make a difference?

In part 2, how many similar and different types of colonies were produced from each portion of your body before bathing? Did certain parts of your body contain more microbes? What happened after bathing? Did the numbers and/or types of colonies change? Did they all change in the same way, or did some areas show different amounts of change?

Going further

- In both parts of this project, you dried your hands with a bath towel as you normally would. Is it possible that you were recontaminating your hands with the towel? Modify the experiment to see if the towel is an important factor.
- You can also continue the project by staining the cells you've collected and observe them under the oil-immersion objective. (See chapter 13 for information on gram staining.) Try to identify the types of bacteria by looking at the results of staining (for example, gram positive or negative) and their shapes. Do different types of bacteria live on different parts of you?
- Compare the difference between a shower and a bath. Does one get you cleaner than the other? Compare the different microbes found on your skin in different types of weather, for example, on a hot, dry day versus a hot, humid day. Does weather affect the numbers of microbes?

Suggested research

- Read about studies of commonly used products such as deodorants that claim to control microbes on our bodies. Look at material from independent sources, such as universities, and also from the companies that manufacture the products.
- Read about studies that investigated "real world" ways in which pathogens are spread among members of the same household.

16

Soils' microbes

Do different kinds of soils contain different numbers & types of microbes?

Background

Organic matter is recycled in the soil. This recycling occurs when dead plants and animals are converted by bacteria, fungi, and other organisms into compounds that live plants can use to grow. These soil microbes are essential players in recycling nutrients and ensuring continued plant growth. Do different types of soils contain different numbers and types of microbes?

Project overview

To determine the number of microbes in soil, you must use a technique that will produce a culture plate containing between 30 and 100 colonies. If you try to count bacteria with fewer than 30 colonies, random chance will cause a lot of errors. If you try to count a plate containing more than 100 colonies, they will be too close together to get accurate readings.

To find a dilution of soil that will produce between 30 and 100 colonies, you will use a *quantitative plating method*, which creates a series of plates containing varying dilutions of soil. You will then compare the microbes found in different soil types.

Materials list

- 4 pint-sized mason jars, with lids
- 100-ml graduated cylinder (with 1 ml markings)
- Sterile water (available from a pharmacy)
- Peat (soil)
- Potting soil
- Compost (soil)
- Marker
- 15 sterile pipettes (with 0.1 ml and 1 ml markings)
- 18 sterile nutrient agar plates
- Clock
- Autoclave (optional)
- Incubator (optional)

Procedures

Use an autoclave to sterilize four mason jars. (If you don't have an autoclave, use the alternative methods for sterilization explained in chapter 1.) Label these jars *peat 1, peat 10^{-2}, peat 10^{-4},* and peat *10^{-6}.* Add 100 ml of sterile water to jar 1 and 99 ml each to the remaining three jars, as shown in Fig. 16-1. Next, add 1 teaspoon of peat to jar 1. Put the lid on the jar and vigorously shake it. This jar contains one teaspoon of soil to 100 ml of water. Let this jar stand for 15 minutes and shake once again.

Use a sterile pipette to transfer 1 ml of the liquid from jar 1 to jar 10^{-2}. Shake this second jar very well. The second jar now contains 100 times less soil (1:100 ratio) than the first jar. Next, use a sterile pipette to transfer 1 ml from the 10^{-2} jar to the 10^{-4} jar and shake the jar vigorously. Repeat this 1 ml transfer from the 10^{-4} jar to the 10^{-6} jar, using another sterile pipette and shaking the jar vigorously. The final jar contains a solution that has been diluted one million times that of the original solution.

These four jars will be used to prepare eight culture plates of varying dilutions. Label eight agar plates as follows: *peat 1, peat 10^{-1}, peat 10^{-2}, peat 10^{-3}, peat 10^{-4}, peat 10^{-5}, peat 10^{-6},* and *peat 10^{-7}.* Use a sterile pipette to transfer 1 ml from jar 1 onto plate 1. Tilt the plate to spread the liquid evenly over the plate's surface.

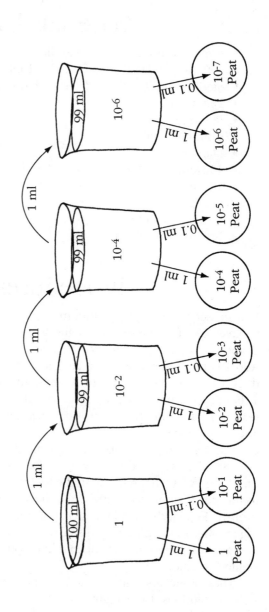

16-1
The teaspoon of soil will be placed in the first of four dilution beakers. Each of the four beakers will be used to produce two petri dish dilutions.

Using the same pipette, put 0.1 ml from the same jar (1) on plate 10^{-1} and tilt to spread. Since you took 0.1 ml instead of 1 ml from jar 1, the plate 10^{-1} contains 100 times less soil than plate 1.

Use another sterile pipette to put 1 ml from jar 10^{-2} onto plate 10^{-2} and spread. Then, use the same pipette to put 0.1 ml from the same jar onto the 10^{-3} plate and spread. Put 1 ml from the 10^{-4} jar onto the 10^{-4} plate, and 0.1 ml onto the 10^{-5} plate, and spread. Put 1 ml from the 10^{-6} jar onto the 10^{-6} plate, and 0.1 ml onto the 10^{-7} plate, and spread. Seal all the plates with adhesive tape.

You have now completed creating an entire series of plates for the peat, with dilutions from 1 all the way down to 1 to 10^{-7}, which is 1:100,000,000. Remember the goal is to find a dilution that will create between 30 and 100 colonies, the perfect number to count.

Dispose of the jar contents and re-sterilize the jars and pipettes. Now repeat this entire process for another two types of soil: compost and potting soil. Once you have created three complete sets of plates, incubate all the plates at 37°C for 24 to 48 hours. Then, count the number of colonies on each plate for all the soil types. The most concentrated plates (the 1's) should have the most growth. Use the dilution that produces between 30 to 100 colonies per plate.

You can now calculate how many bacteria were found in each soil type. To obtain the number of bacteria per ml of water in jar 1 (the number of bacteria per 0.01 teaspoon of soil), divide the number of colonies on the plate by the plate dilution. For example, if there are 65 colonies on plate *compost 10^{-3}*, then 65 colonies divided by 10^{-3} (−1,000) = 6500 bacteria. This would mean that there was 6500 bacteria per 0.01 teaspoon of soil placed in the first jar labeled 1. (If there are 6500 in 0.01 teaspoon of soil, you can determine there were approximately 650,000 bacteria in the teaspoon of the peat soil that was placed in the first jar at the beginning of this project.) Repeat these calculations for each type of soil.

Analysis

Compare the numbers of bacteria growing in each type of soil. Are there different numbers of colonies in each soil type? Do different types of soils contain significant differences in number of microbes? Of what importance are these numbers?

Going further

- Do this experiment with different types of media, such as Sabouraud or tryptic soy. Do you get different results with the same soil samples?

- Run this experiment, but concentrate on the bacteria found.

Suggested Research

- Research the importance of your findings. How can this information be applied both in your garden and on the farm?
- Read about the importance of these soil microbes and the role they play in the soil ecosystems.
- Can soil microbes be used to quantify the health of a soil? Has anyone developed a scale of this type?

Part 6

Controlling microbes with antimicrobial products

This section contains three projects that investigates the use of manmade antimicrobial products including paints and common household disinfectants like Lysol. It also looks into the new phenomenon of "super germs."

Have you read about new strains of bacteria that don't respond to antibiotics? These new bacteria have become resistant to traditional antibiotics. The first project in this section allows you to investigate this phenomenon. Is it possible for the common bacteria found in your bathroom to become resistant to common household disinfectants? How long would it take for this to occur?

People who own boats know that they are in a constant battle with microbes and other organisms that attach themselves to the hull of a boat. To prevent this biofilm from growing, they use one of many anti-fouling paints that prevents this growth. Since most of the traditional anti-fouling paints are highly toxic to marine life, a new type of paint has been developed that is supposed to stop the growth but be safe for the marine ecosystem. Do such paints work as well as the traditional paints? The second project in this section will help you find out.

The third project in this section investigates the usefulness of our traditional methods of disinfecting things in our homes. How effective are disinfectant sprays and dishwashing detergents at reducing bacteria?

17

Resistant bacteria

Do bacteria become resistant to household disinfectants?

Background

Have you ever read about "super germs"? Most bacteria are killed by the antibiotics prescribed by your doctor, but super germs are new strains of these bacteria that have become resistant to many common antibiotics. An illness that once took a week of antibiotic pills to cure might be unaffected if it is caused by one of these super strains of bacteria.

Many scientists and doctors believe that these super germs have become resistant because of decades of antibiotic overuse. Gradual genetic changes (mutations) have altered the makeup of the bacteria, producing strains that are resistant to the antibiotics that once killed them.

This is a classic case of natural selection. Those bacteria poorly suited to fight the antibiotics died, while those more resistant to the antibiotics survived, passing on their stronger genes to future generations of bacteria, creating a "super race" of germs.

Seeing changes occur in a species is usually difficult because numerous generations must pass for small changes to become visible. Bacteria reproduce rapidly enough for these changes to be noticeable.

Project overview

If bacteria become resistant to antibiotics prescribed by your doctor, can they become resistant to common products used to reduce or eliminate them in your home? Will bacteria become resistant to antiseptics and germicides such as iodine or Lysol? If they do become resistant, how long does it take?

In this project, you will culture some common bacteria and create antibiotic discs soaked in various disinfectant products. Those bacteria growing most closely to the discs (the most resistant) will be collected and cultured. This process will be repeated five times. The resulting colonies of bacteria will be cultured and tested once again to see how resistant they have become to the disinfectant over this period.

This project is divided into six sections to make it easier to follow.

Materials list

- Autoclave
- Blood agar plates (available from a supply house)
- Marker
- Sterile cotton swabs
- Cultures of *E. coli* on nutrient agar plates (available from a supply house)
- Cultures of *Micrococcus luteus* on nutrient agar plates (available from a scientific supply house)
- Cultures of *Serratia marcescens* on nutrient agar plates (available from a supply house)
- Small piece of thin glass tubing (about 0.3 cm in diameter)
- Graduated pipettes
- Antiseptics (you can add your own selections):
  ~Hydrogen peroxide
  ~70% rubbing alcohol
  ~Iodine
- Germicides (you can add your own selections):
  ~Lysol (liquid, not spray)
  ~Hilex (liquid, not spray)
  ~Dispatch (a hospital disinfectant)
- Incubator
- Vernier calipers
- Inoculating loop
- Sterile nutrient broth (available from a supply house)
- Flask

- Test tubes
- Test tube caps
- Test tube rack
- Sterile pipette
- Spreading rod

Procedures

This project requires that all glassware be sterilized in an autoclave prior to use. Re-sterilize inoculating loops and spreading rods before and after each use. Be sure to keep your table surface clean and follow sterile procedures as described in chapter 1 throughout this project.

 Caution! Sterilizing spreading rods is a dangerous procedure that must be performed by your sponsor. See chapter 1 for more information on sterilization techniques.

Preparing the plates

You will first inoculate the plates with bacteria and then create "wells" to hold the disinfectants. Prepare one blood agar plate for each disinfectant to be tested. Divide each plate in half by drawing a line on the lid with a marker. Label one half *zone A* and the other *zone B*.

Gently pass a sterile cotton swab over the *E. coli* culture plate. Use this swab to inoculate the blood agar plate, making a series of horizontal streaks across zone A as shown in Fig. 17-1 and a series of vertical streaks across zone B. This streaking method makes it easier to identify the zones. Be sure each plate is marked with the name of the bacteria.

Next, create "wells" in the agar to hold the disinfectants. Use a thin, hollow piece of sterile glass tubing to gouge out a well (hole) about 0.3 cm in diameter in the middle of zone A, and another in the middle of zone B. Use a sterile graduated pipette to fill each well with 0.1 ml of the first disinfectant. Tape the plates shut.

Repeat this procedure for each type of bacteria and each type of disinfectant. There will be one plate for each disinfectant tested for each bacteria, so if you are testing three types of bacteria on six types of disinfectants, you will have 18 blood agar plates (3 times 6). Label all the plates with the name of the bacteria and the disinfectant in the wells.

Incubating & taking readings

Incubate all the plates for 24 hours at 32°C. After the incubation period, remove the plates from the incubator and measure the zones of inhibition around each well with a vernier calipers. The zone of inhi-

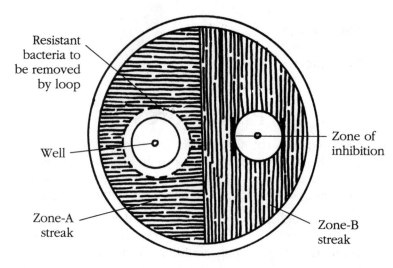

Resistant bacteria to be removed by loop

Well

Zone-A streak

Zone of inhibition

Zone-B streak

17-1 *The petri dish is divided into zones A and B. Each zone has a well containing the disinfectant. The colonies that grow closest to the disinfectants will be cultured.*

bition is the ring around the well where the bacteria doesn't grow. Record the width of this zone.

Selecting & culturing cells

Next, remove the colonies of bacteria closest to the zone of inhibition around the zone-A well with a sterile inoculating loop. This bacteria is used to inoculate a new plate. Use the loop to streak the zone-A half of the new plate. Use the bacteria removed from zone B to streak the zone-B half of the new plate. After streaking the plate, use a sterile glass rod to make wells on each streak as you did on the first set of plates. Fill the wells with the same disinfectant used in the previous plate. Repeat this procedure for each type of bacteria and each type of disinfectant.

Incubate this second group of plates for 24 hours at 32°C. Repeat this procedure of selecting and culturing cells five times over a five-day period, measuring the zone of inhibition each time. Continue to record your data. Those cells showing the most resistance to the disinfectant are continually selected used to inoculate the next set of plates.

Isolating the resistant cells

Collect cells from the final set of plates with a sterile inoculating loop. Use this sample to streak a set of sterile nutrient agar plates as shown

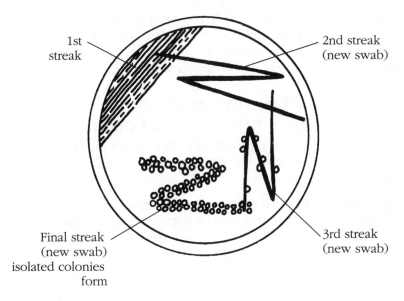

1st
streak

2nd streak
(new swab)

Final streak
(new swab)
isolated colonies
form

3rd streak
(new swab)

17-2 *To create isolated colonies, follow these four steps of streaking.*

in Fig. 17-2. Use a quadrant-streak method to isolate the resistant colonies by following the four steps shown in the figure.

Incubate these plates for 24 hours at 32°C. After incubation, remove the plates. Use a sterile inoculating loop to remove four colonies at least 3 mm in diameter, isolated, as shown in the figure. Use the inoculating loop to place these cells in a sterile flask containing 40 ml of sterile nutrient broth. Incubate this broth at 32°C for three hours to reach optimal growth. This flask now contains a rich culture of the most resistant cells from the previous steps. Label each flask *final/* followed by the name of the bacteria and the name of the disinfectant, for example, *final/E. coli/Lysol.*

Diluting the final culture

After incubation is complete, remove the flask from the incubator. Since this culture is so rich in bacterial cells, you must dilute it before continuing. You will dilute this culture by a factor of 1,000,000. To create the dilution, add 9 ml of sterile nutrient broth to each of six sterile test tubes. Use a sterile graduated pipette to remove 1 ml from one of the *final/* flasks created in the previous step. Transfer this 1 ml to the first sterile test tube and mix well. Label this first test tube *10* since it is diluted to a 1:10 ratio.

Next, use a sterile pipette to take 1 ml from this tube (10) and transfer it to the second tube. Label this tube *100*, and mix well. Take

1 ml out of the 100 tube and transfer it into the third tube, and so forth, until there is a dilution of 1:1,000,000 (six tubes). The final dilution will be used to inoculate the final plates. This process must be repeated for each bacteria and disinfectant tested.

Creating, incubating, & counting the final plates

Use a sterile graduated pipette to place 0.1 ml of a disinfectant on the surface of a sterile blood agar plate. Spread it evenly around the plate with a sterile spreading rod.

 Caution! Remember, sterilizing spreading rods is a dangerous procedure that must be performed by your sponsor. Again, see chapter 1 for more information on sterilization techniques.

Use a sterile pipette to transfer 0.1 ml of the diluted bacteria from the 1,000,000 tube (created in the previous step) to the same plate, placing it on top of the disinfectant. Once again, spread evenly with a sterile spreading rod. Incubate these plates at 32°C for 24 hours. Repeat this process for each bacteria and disinfectant tested. Inoculate two additional plates with bacteria, but add no disinfectant. These will be your controls.

After incubation is complete, count the number of colonies growing directly on the disinfectant. Each colony can be counted as one bacterial cell. The number of cells capable of growing on the disinfectant indicates the degree of resistance established over the five-day period of this project. Discuss with your sponsor how to properly dispose of all the cultures used in this project.

Analysis

Did all or any of the bacteria show at least some resistance toward all or any of the disinfectants? Fill in a table similar to the one in Fig. 17-3. Did the bacteria show more resistance toward the antiseptics or the germicides?

Did the bacteria that showed the strongest inhibition to a bacteria in the beginning of the project also show the most resistance at the end of the project? Rank the order of the disinfectants' ability to prevent resistance for each bacteria. Did any bacteria become 100% resistant to any disinfectant?

BACTERIA TESTED	TRIAL#	ZONE	ZONE OF INHIBITION IN MM DAY #					AVG% REDUCTION (DAY 5/DAY 1)
			1	2	3	4	5	
E. coli	1							
	2							
	3							

DISINFECTANT

17-3 *Record your data in a table similar to this one.*

Going further

- Run a similar experiment, but concentrate on microbes other than bacteria, such as molds.
- Devise an experiment to see how this project applies to everyday life around the home. Would it make sense to change disinfectants every few weeks?

Suggested research

- Research microbial resistance and its importance to us.
- Look into the similarities between this project and the process of evolution.
- Read the latest literature in medical journals about super germs.

18

Biofilms & biofouling

Comparing toxic to environmentally safe anti-fouling paints

Background

Microbes in moist or wet habitats often adhere (stick) to a surface, forming a living sheet or film. This mass of microbes attached to a surface is known as a *biofilm*. The types of microbes that make up this biofilm vary depending upon the habitat on which they grow. Some microbes, such as bacteria, can grow without any light, while others, such as algae cells, require light.

A biofilm comprised mainly of bacteria is responsible for the uncomfortable feeling we have on our teeth and gums when we wake up in the morning. A biofilm comprised mainly of algae often grows on the sides of home aquariums.

Almost any moist surface can produce biofilm. When biofilms cause problems, the process is called *biofouling*. For example, ships can have problems moving through water because of the biofilm and other organisms that feed on this film growing on their hulls. Biofouling also occurs in water-treatment plants and other industries that use water. There are different kinds of chemical substances used to control biofilm and prevent biofouling.

Project overview

The boating industry is the biggest user of anti-biofouling products. Commercial ships, military vessels, cruise ships, and pleasure craft all require some method of keeping biofilms from growing on their hulls. (See Fig. 18-1.) The biofilm becomes food for other organisms, some of which also attach themselves to the hulls of boats, interfering with proper handling and efficiency.

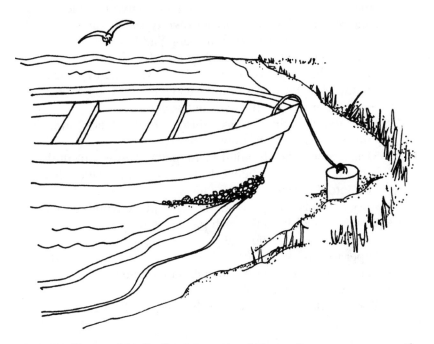

18-1 *Biofilms and biofouling occurs in places as diverse as your mouth and the hulls of boats, as shown here.*

Unfortunately, many of the substances used to prevent biofouling not only kill the organisms that create biofilm, but many types of marine plants and animals as well. Some of these anti-fouling products have been banned, and others have restricted uses. A new generation of anti-fouling paints is being sold with hopes that it can do its job without harming the environment. How do different types of anti-fouling paints compare in how they reduce or eliminate biofilm growth?

Materials list

- About 20 pint mason jars
- Pond water
- A few desk lamps
- Marker
- A selection of anti-fouling marine paints from a marina supply store (You must read the labels carefully to determine which additives are found in the paints. You need a very small amount of each. Use white paint only.) Use the suggestions below or find your own types of paints:
  ~Oil-based paint with an anti-fouling additive
  ~Latex paint with an anti-fouling additive
  ~Anti-fouling paint containing TBT (TBT is highly toxic to marine life if it gets into the water)
  ~Anti-fouling paint that does not contain TBT
  ~Anti-fouling paint that uses silicon or other "slick" surface instead of toxicity to prevent biofilm
  ~Hard "non-ablative" anti-fouling paint (This is a paint that may contain toxic chemicals, but is not supposed to leach off into the water.)
- About 10 small, disposable, model airplane paintbrushes
- 2 microscope slides for each paint tested
- Coverslips for the slides
- Paper towels
- Microscope with an oil-immersion objective
- Stereoscope (optional)
- Book to identify aquatic microbes (especially algae)

Procedures

Fill each mason jar ¾ full with pond water. (The jars need not be sterile.) Leave all the jars under a desk lamp for seven days. Don't keep the lamp too close to the jars or the water will overheat, killing the microbes. Add more water if the water level drops.

After seven days, there should be a rich growth of microbes in the mason jars. If the jars are not greenish in color, add more pond or aquarium water and wait another few days. You must see algae growth (the water should have a greenish tinge) in or on the mason jars before continuing.

Caution! The paints used in this experiment give off potent and dangerous fumes. Do all work outdoors or under a ventilation hood. Store any paint left over from the experiment according to the directions on the cans.

Using the first small paintbrush, apply a thin coat of the first paint to one side of a microscope slide as shown in Fig. 18-2. Paint two slides for each paint tested. Label the slides appropriately. Use a different brush for each paint to be tested. Let all the slides air-dry for two days. Be sure to keep the slides under a ventilation hood, outdoors, or in a well-ventilated room.

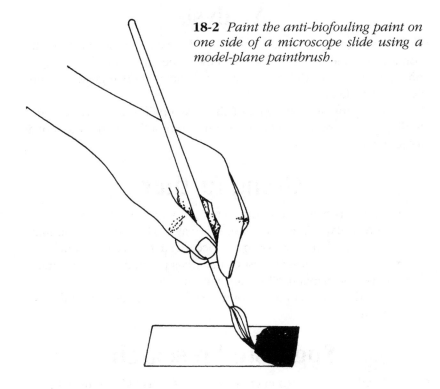

18-2 *Paint the anti-biofouling paint on one side of a microscope slide using a model-plane paintbrush.*

Carefully place the two similar slides in one of the prepared mason jars. The painted sides of the slides should face up in the water. Label each jar with the paint used on the two slides. Put an extra two unpainted slides in another jar to act as a control. Leave all the slides in the jars for 14 days. During this period, add water to the jars to keep the slides underwater. Keep the jars under the lamps or in a sunny room.

At the end of two weeks, remove all the slides. Be sure the labels are still visible on each slide. Wipe the uncoated side of the slides clean with paper towels. Observe the slides under a microscope, at 100X and 400X. (You could also look at the slides with a stereoscope.) Take notes on the number of microbes found on each slide. You'll have to increase the amount of light passing through the condenser of the scope, since the slides have paint on them. You might even need an additional light source.

Finally, put a drop of water on the center of each slide and cover the drop with a coverslip. View each slide under oil immersion. Make notes on the types and number of organisms found on each slide, including the controls.

Analysis

Did biofilm grow on all the slides? Was there a difference between the controls and the painted slides? Did the same amount of biofilm grow on all the slides, or did some of the treatments reduce growth or prevent it completely?

Does it appear that some of the new environmentally safe antifouling paints are as good at preventing biofouling as the traditional, toxic paints?

Going further

- This experiment was run in the lab with fresh water. Design a similar experiment, but use salt water. This can be done in a saltwater aquarium or in the ocean, such as along a dock.
- Devise a small-scale test of materials painted with the same types of paints used in this experiment.
- Create an experiment around biofilms that you can find in your home.

Suggested research

- Read about environmental problems caused by anti-fouling paints and other products.
- Contact environmental organizations such as the Center for Marine Conservation in Washington, D.C., for more information about this subject.
- Speak with recreational boaters to see how they feel about environmental concerns involving anti-fouling paints.

19

Microbes in the home

How effectively do we eliminate microbes from our homes?

Background

Microbes are all around us, including the places where people try the hardest to remove them. They are on the utensils we try so hard to clean with dishwashing detergents, and all around our bathrooms where we spray disinfectants to kill germs.

When we clean areas in the bathroom with disinfectant sprays and wash dirty dishes in a dishwasher or by hand, we are fighting an ongoing battle to remove unwanted microbes from our homes. How successful are these cleaning procedures and products?

Project overview

All surfaces in our homes contain at least some microbes. Some of them might be disease-causing, but most are not. Bathrooms are a room in which microbes can thrive. Household disinfectants are used to reduce or eliminate microorganisms that live with us in the bathroom.

Dirty dishes are also an excellent place for microbes to flourish. The leftover food that sticks to utensils and plates provides food and a home for many microbes. Dish detergents are supposed to clean off not only the remaining food, but the microbes living in the food.

This project is divided into two parts. The first investigates how effectively disinfectants remove microbes from surfaces in the bathroom. The second part determines how effective dish detergents are at cleaning the microbes off of dirty utensils.

Materials list
For part 1

- At least 6 sterile nutrient agar plates
- Sterile cotton swabs
- Sterile antibiotic filter-paper discs
- At least 3 different disinfectant sprays, such as Lysol
- Sterile forceps
- Incubator

For part 2

- 2 dirty forks that have been used for eating a meal within the past hour
- 2 clean forks that have been washed by hand and air-dried after being used for a meal
- 2 clean forks that have been washed in a dishwasher after being used for a meal (optional)
- 6 sterile nutrient agar plates
- Thermometer

Procedures
Part 1

You must use sterile (aseptic) technique throughout this procedure (see chapter 1 for details). Use a sterile cotton swab to wipe the surface of your bathroom sink, bathtub, or any other surface that is typically cleaned with a disinfectant. Take the contaminated swab and wipe it across a sterile nutrient agar plate to inoculate the plate with the microbes found on the bathroom surface. Repeat this procedure with all six plates, using the same technique to create all six plates.

Label the plates with the source of the sample, for example, *around sink.*

Prepare the antibiotic filter-paper discs by spraying six discs with one of the disinfectants until they are damp but not dripping. Use sterile forceps to place one disc into each plate. Label that sector of each plate with the name of this disinfectant.

Repeat this procedure with the other two types of disinfectants. Place each disc equidistant from the others, as shown in Fig. 19-1. You'll end up with six plates, each containing three discs from three different types of disinfectants. Tape all the plates closed with adhesive tape and incubate at 37°C for 24 to 48 hours. After this period, compare the zones of inhibition found around each disc. Do this through the plate cover, without opening the plates.

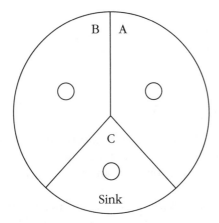

19-1
Place three discs, each containing a different type of disinfectant, equidistant from each other on an agar plate.

Part 2

Label two nutrient agar plates *clean/handwash,* two others *clean/ dishwash* and the last two *dirty.* Take a dirty fork and press the tines into the agar surface of the *dirty* plate. (See Fig. 19-2.) Repeat this for the other dirty fork, using the other plate. Repeat the same procedure with the two *clean/handwash* forks and again with the two *clean/dishwasher* forks. You'll end up with three sets of two plates. Seal all the plates with adhesive tape.

Place all the plates in an incubator at 37°C. Incubate the plates for 48 hours. Observe and make notes of the number, placement, and shape of the colonies on the plates.

19-2
Insert the tines of the fork into the agar.

Analysis

For part 1, did any of the disinfectants do a better job of controlling the bathroom microbes than the others? Does it appear that disinfectants eliminate or reduce the number of bacteria and other microbes on surfaces in your bathroom?

For part 2, are there differences between the clean forks and the dirty forks? Are there differences between the two types of clean forks? What can you conclude about bacteria on the surfaces of so-called clean utensils? Does a dishwasher make a difference in the number of microbes found on the fork?

Going further

- For part 1, use different disinfectants. Are they all as efficient or inefficient?
- Do part 1 in different areas of the house. Do you get different results?
- Repeat part 2 with forks cleaned in the dishwasher using different types of dish detergents. Are there any differences?

- For part 2, compare the growth found from a clean fork over different periods of time since it was cleaned. How rapidly do microbes accumulate on the forks?

Suggested research

- Research the active ingredients of disinfectants, antimicrobial products and dish detergents. How do these products work? Do any have antibiotics in them?
- See if there is a direct relationship between the cost of products tested and their effectiveness.

Part 7

Microbes & humans

The following four projects show four different ways in which people affect microbes and microbes affect people. The first project is about a relatively recent environmental concern: the effect of electromagnetic radiation on our health. This project investigates whether this common form of radiation has an effect on a terrestrial microscopic organism, the nematode.

The next project studies buffers. When people use microbes to produce products such as wine, beer, cheese, and yogurt, the microbes must be artificially maintained. This can cause problems. Buffers are substances used to resolve these problems. The third project looks at one of the classic methods of identifying and classifying bacteria, gram staining.

The final project returns to the subject of electromagnetic radiation. This project, however, tries to determine whether this form of radiation is harmful to aquatic organisms.

20

Electromagnetic fields & soil organisms

The effects of electromagnetic radiation on nematodes

Background

Anything connected to an electrical circuit has an electric field and a magnetic field surrounding it. For example, a personal computer and a high-tension wire are both surrounded by these fields. Since this form of radiation consists of both electric and magnetic fields, it is called *electromagnetic radiation*.

Studies about the effect of this type of radiation on people are inconclusive. Some studies have shown no ill effects, while others have shown increased risks of diseases including some forms of cancer from continued exposure to this type of radiation.

Project overview

Although most of our concerns about electromagnetic radiation are about the health risks to people, what about its effect on other organisms? Does it affect soil organisms? If so, it can have ramifications throughout the local ecosystem. In addition, concerns about our health might be heightened if we (knew) that other organisms were harmed by this radiation. If it harms other organisms, might it affect us?

This project investigates whether the most common soil organism, the nematode (a small roundworm), is affected by electromagnetic radiation. Since these organisms are important to most terrestrial ecosystems, a negative effect on these organisms would be felt throughout the ecosystem.

This project is divided into two sections. In the first, you collect soil nematodes from some local topsoil to be used as test specimens. In the second part, you will see if electromagnetic radiation has any effect on these nematodes.

Materials list

- Nematodes (The "Procedures" section describes how to collect your own, but they can be purchased from a scientific supply house.)
- A few paper lunch bags
- 2 funnels (with mouths slightly larger than the mouths of the beakers)
- 2 pieces of rubber tubing (approximately 6 to 8 inches long with a diameter to fit the end of the funnels)
- 2 pinch clamps
- 2 rings and ring stands
- Paper tissue
- Rubber bands
- Petri dishes
- 2 beakers
- Three 6-volt batteries (from a hardware stove)
- Wire (long enough to be coiled several times around the beaker)
- Microscope (low power is sufficient)

Procedures
Part 1

The following instructions show you how to catch your own nematodes using the *Baermann Funnel Extraction technique*. The best place to look for nematodes is in soil that naturally retains moisture and has an abundant amount of grass growing on it. Fill a few small lunch bags with this soil, from the top four inches of soil.

To create a Baermann funnel, start by attaching one end of the rubber tubing to the end of the funnel as in Fig. 20-1. Close the other end of the tubing with the pinch clamp as shown. Place the funnel on

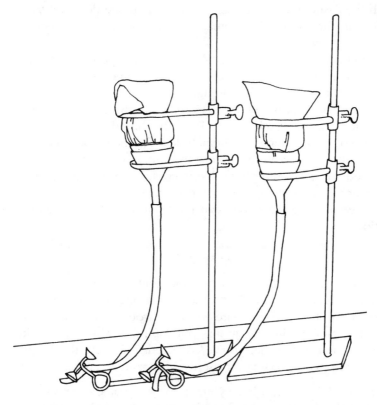

20-1 *The jar containing the soil is covered with tissue paper and placed upside down over the funnel. The funnel leads to tubing that is closed with a pinch clamp at its end.*

the ring in the ring stand with the tubing passing through the ring, as shown in the figure. Now completely fill the tubing and the funnel with water. You might have to let some of the water through the pinch clamp to remove air bubbles.

Next you'll prepare the soil containing the nematodes. Crumble one full tablespoon of the soil you've collected and place it in a beaker. Fill the beaker with water about ¾ full. Carefully place a tissue over the mouth of the beaker (containing the soil and water) and secure it with a rubber band. The tissue should be taut, but don't let it tear.

Now slowly and carefully invert the beaker and place it upside down on top of the funnel, as shown in Fig. 20-1. The tissue will sag enough so it comes in contact with the water in the funnel. This is important because the nematodes will pass from the beaker water through the tissue into the funnel water. If the tissue does not touch the water in the funnel, the nematodes do not have a continuous medium in which to travel.

The nematodes will slowly settle on the tissue, permeate through the tissue, settle through the water in the funnel and finally collect in the water at the bottom of the tubing, just above the pinch clamp.

After 24 hours, draw off 10 to 15 ml of the water from the tubing by releasing the pinch clamp. Pour this liquid into a few petri dishes so you can observe the nematodes. The nematodes can survive off of the bacteria in the water for about two weeks if they are refrigerated. Your soil sample probably contains nematodes that are plant parasites, predators, and microbe grazers. Now that you've collected your nematodes, you can move on to the actual experimentation.

Part 2

You'll subject the nematodes to two different strengths of electromagnetic fields. The field will be created by using one or two 6-volt batteries connected to coiled wire wrapped around the nematodes.

For the first experimental group, put 200 ml of water in a beaker and add 30 nematodes that you've collected. To create the electromagnetic field, wire two 6-volt batteries in parallel to create 12 volts.

 Caution! Be sure to have your sponsor or some other individual who is knowledgeable about electrical circuits check your wiring before proceeding.

Attach the wire to the batteries. Then wrap the wire around the beaker containing the nematodes, at least 10 times. Leave the beaker with the nematodes in place for five hours.

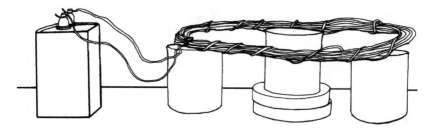

20-2 *The electromagnetic radiation is supplied from the coils that surround the beaker.*

Create a second experimental group the same as the first, but using only one 6-volt battery as in Fig. 20-2. Create a third group, but use no batteries and be sure it is well away from any electric devices. This will be the control group. Leave these two groups in place for five hours, also.

After five hours have elapsed for each group, use the low power on your microscope to observe the nematodes. Separate the live nematodes from the dead ones and calculate survival rates for each group. (Live nematodes are usually in constant curling motion, while the dead ones are nearly straight and immobile.) You can check the nonmoving nematodes by touching them with a fine probe to see if you get a response. If they do react, they are probably near death (classify them as "ill"). Categorize the nematodes as healthy, ill, or dead.

Analysis

Plot the number of live, ill, and dead nematodes for each group on a graph similar to Fig. 20-3. Was there a difference between the control and the two experiment groups? Does it appear that the radiation is harmful to nematodes? Was there a difference between the two experimental groups? Does the level of radiation affect the mortality rate of the nematodes?

Going further

- This project uses terrestrial nematodes, but tests them in water. Devise an experiment that would test nematodes in their natural habitat, the soil. You can either do this in the lab

Survival rate

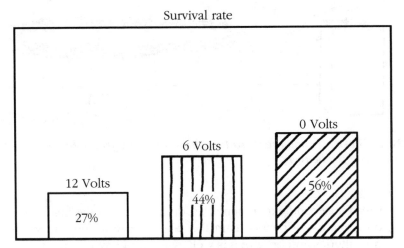

Percentage of living nematodes

20-3 *Create a graph that represents the differences between the control, 6-volt, and 12-volt groups.*

by using petri dishes containing a thin layer of soil, or in the field by sampling nematodes under high-tension wires and in other areas.

• Enhance this project by renting or borrowing a gaussmeter that reads the amount of electromagnetic radiation present. Measure the actual amount of radiation that the nematodes were subjected to.

• This project uses direct current (DC) created by batteries. Household current, however, uses alternating current (AC). Modify this project to use AC to more accurately simulate EMR fields.

Suggested research

• Research the latest findings about electromagnetic radiation and health risks. Be sure to use the latest journals and reports, since new information appears each month.

• Besides looking into the harmful effects of this radiation on people, look into how it might affect local ecosystems.

21

Buffers

How important are buffers to the survival of bacteria?

Background

When an organism lives in a habitat, it has a tendency to change that habitat. For example, sun-loving trees grow in areas with lots of sun, but once they reach maturity, they make it impossible for more sun-loving trees to grow since they themselves are blocking the sun. Only shade-tolerant trees can then grow.

Bacteria, like other organisms, need the right environment to grow and reproduce. And like other organisms, they tend to change their environment. In this case, it is often their waste-products that can affect their ability to continue to grow and reproduce. Wastes can build up, creating a concentration of toxins too high for the bacteria to continue to grow. Waste and other byproducts of metabolism can change the pH of the bacteria's habitat, making it unsuitable for them to continue living in their environment. (See Fig. 21-1.)

Project overview

Scientists who study microbes and those responsible for managing microbes in industry (such as in breweries or some dairy-product production) must constantly be concerned about changes in pH

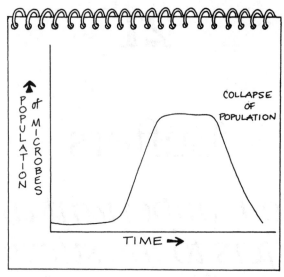

21-1 *Create a graph to represent the population of microbes over time.*

caused by the very microbes that they are studying or culturing for food production. The byproducts these microbes produce cause their environment's pH to change—sometimes to the point where the microbes can no longer survive.

Scientists and industry use compounds called *buffers* to control changes in pH. Their purpose is to maintain a constant pH in the microbe's environment. How important are buffers in maintaining the pH of a culture? What is the effect of a buffer on bacterial growth?

Materials list

- Two 250-ml flasks
- 2 grams of tryptone (available from a supply house)
- 2 grams of yeast extract (available from a supply house)
- 2 grams of glucose (available from a supply house)
- Balance
- Spatula
- Marker
- 0.5 grams K_2HPO_4 (potassium phosphate dibasic) to act as a buffer (available from a supply house)
- 100-ml graduated cylinder
- 2 stirring rods
- Sterilization filter syringe (0.45 micron or smaller)
- 6 test tubes with screw caps

- Sterile inoculating loop
- Bunsen burner (optional if using disposable loops)
- Culture of *Streptococcus faecalis* (available from a scientific supply house)
- Incubator
- Turbidity comparison chart (available from a scientific supply house)

Procedures

You must use sterile technique throughout this project. (For details about sterilization, see chapter 1.) Place into each of two flasks the following: one gram of tryptone, one gram of yeast extract, and one gram of glucose. This provides an environment for the bacteria to live. Label one of the flasks *with buffer* and the other *no buffer*. Next, add 0.5 grams of K_2HPO_4 (potassium phosphate dibasic) into the flask labeled *with buffer*. Add 100 ml of water to each flask and stir until dissolved.

Now you will sterilize the solutions in both jars using a filter sterilization syringe. This syringe sterilizes the liquid by filtering out all organisms. To do this, fill a sterile syringe with the *no buffer* solution. Put the sterile filter at the end of the syringe, and force the liquid through the filter. Place the liquid into a sterile test tube labeled *no buffer*. Cap the tube when it contains 10 ml of the media. Fill two more *no buffer* test tubes with 10 ml of media. Cap each tube.

Repeat this procedure using the *with buffer* media. Filter sterilize this media by passing it through the sterile filter, and put 10 ml into three *with buffer* test tubes.

Inoculate each of the six tubes using a sterile inoculating loop with the *Streptococcus faecalis* culture. A single loopful is all that is needed. (See chapter 1 for instructions on inoculating.)

Observe the initial *turbidity* (cloudiness) of the tubes. This will be turbidity = 0. Incubate all the tubes at 37°C for 48 hours and then observe the degree of turbidity again. Scale the turbidity level from 0 to 5, with 0 being no turbidity (initial conditions), 1 being very slight cloudiness, and 5 being an extremely cloudy, dense culture.

The turbidity gives an estimate of the microbial growth in each tube. The higher the number, the more growth has occurred. Use a turbidity comparison chart to make your observations more exact. (See Fig. 21-2.) There are also more sophisticated devices to make bacterial counts. Speak with your sponsor to see if any of these devices are available at your school.

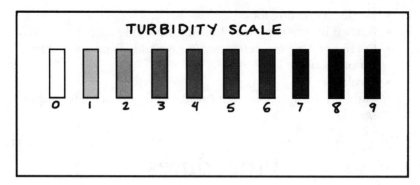

21-2 *A turbidity chart is a simple way to compare the amount of microbe growth.*

Analysis

How did the buffered and non-buffered cultures compare? Did the differences appear at once, or did it take a considerable amount of time for changes to occur? If there was more growth in the buffered tubes, what can you conclude about the effect of the buffers in their enclosed environment? Are buffers needed in nature?

Going further

- Run a similar experiment, but use a *Lactobacillus* species. How do the results differ?
- Run a similar project, but run it on a larger scale such as in an aquarium. How do the results differ?
- Continue the project by quantifying the number of cells in the tubes. (See "Cell counts" in chapter 1 for more information.)

Suggested research

- Research exactly what happened to the microbes' environment.
- Do these buffers occur in nature? If so, why? If not, why not?
- Research how buffers are used in industry. Look into breweries, wineries, and manufacturers of cheeses and yogurts.

22

Gram stains

A survey of common bacteria

Background

Studying cells can be difficult, since most cells appear transparent under a microscope. Staining is a method to identify cells by coloring them. Most stains only color certain types of cells or certain organelles within cells. This occurs because each stain will only bond with certain chemical compounds. Stains that color only certain cells, or parts of a cell, are known as *differential stains*.

Project background

There are many kinds of differential stains. The staining procedure most commonly used in microbiology is the *gram stain technique*. Gram staining is often the first step to microbial identification. Gram-positive cells have thick cell walls that don't let alcohol seep through. This results in a different color pattern than gram-negative cells. Gram-negative cells have thinner cell walls that permit alcohol to enter, which washes away any stain that was previously stuck to the cell wall. Therefore, gram-negative cells are "decolorized" and gram positive cells are not. Are most of the commonly found bacteria gram-positive or gram-negative?

In this project, you will practice your staining technique and become familiar with known gram-positive and gram-negative cells and the staining procedure. Once you learn gram staining, you will survey bacteria found in common habitats and determine whether they are gram-positive or gram-negative.

Materials list

- Microscope slides
- Inoculating loop
- Bunsen burner
- A known gram-positive culture such as *Bacillus subtilis, Bacillus cereus, Clostridium botricum,* or *Streptococcus lactis* (available from a scientific supply house)
- A known gram-negative culture such as a *E. coli, Rhizobium leguminosarum, Spirillum volutans,* or *Vibrio fischeri* (also available from a supply house)
- A series of unknown cultures (You can take samples from virtually anywhere by using a sterile cotton swab and applying the swab to nutrient agar plates.)
- Several pieces of old, dead grass
- Pond water
- Eyedroppers (one for each sample)
- Sterile nutrient agar plates
- Gram staining kit (available from a supply house or probably in your school lab) that contains the following:
  ~Crystal violet stain
  ~Iodine solution
  ~Ethyl alcohol (95%)
  ~safranin red stain
- Paper towels
- Microscope with oil-immersion objective
- Timer or watch with secondhand

Procedures

 Caution! Your teacher or sponsor must be present when you perform gram staining. Whenever using flames (such as for sterilization), be sure you are working with your sponsor.

Label one microscope slide *known* + and another *known* –. Sterilize the inoculating loop with the bunsen burner (following instructions in chapter 1). Dip the loop into the first culture (*known* +) and spread it into a thin film on the microscope slide labeled *known* +. One loopful is sufficient. Let the slide air-dry. Then "fix" the microbes to the slide by passing the slide through the flame two or three times, as shown in Fig. 22-1.

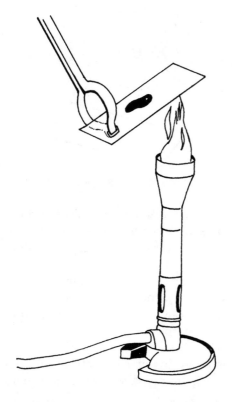

22-1
You "fix" the stain by passing the slide through the bunsen burner flame. Hold the slide with forceps.

Place a few drops of crystal violet stain on the fixed slide, as shown in Fig. 22-2. Let the slide stand for one minute, then gently wash off the excess stain with an indirect stream of water for about two seconds. Next, place a few drops of the iodine solution on the slide and let it stand for three minutes. Wash off the excess as before.

Very gently blot (do not rub) the slide dry with paper toweling. Next, decolorize the slide by adding a few drops of ethyl alcohol to it. Leave the alcohol on for no more than 30 seconds before washing it off and blotting the slide dry. Finally, place a few drops of safranin red stain on the slide for one minute and again wash it off with tap water. Gently blot the slide dry with paper toweling.

You are now ready to view the fixed and stained slide under the microscope using the oil-immersion objective. View the sample. What colors do you see? Gram-positive organisms are purple; they stain

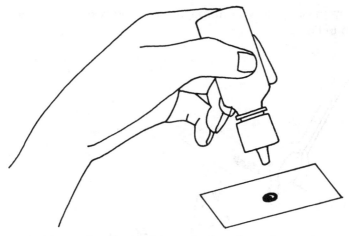

22-2 *Place a few drops of the stain directly on the specimen.*

with the crystal violet and do not decolorize after the alcohol rinse. Gram-negative organisms are pink because they lose their purple color from the crystal violet stain and decolorize after the alcohol rinse.

Repeat this procedure with the known gram-negative culture. You'll know you can do these techniques correctly when your smear from the gram-negative culture comes out pink, and your smear from the gram-positive culture comes out purple.

When you are confident that you can do gram staining correctly, you are ready to survey bacteria found in common habitats and determine whether gram-positive or gram-negative bacteria are more common. First, make a common hay infusion habitat by taking old, dead grass and soaking it in water for one to two weeks. Use a sterile inoculating loop to take a sample of this water and spread it thinly over a microscope slide. Follow the instructions just listed for fixing, preparing, and staining the slide. Repeat this procedure with pond water.

For the other samples, you must take a sample using a sterile cotton swab and inoculate a nutrient agar plate by gently swabbing it across the agar. Use a different sterile cotton swab and plate to take samples from places such as a damp kitchen counter or your socks. Incubate the agar plates for 48 hours at 37°C.

After incubating the plates, follow the instructions given earlier to create a slide and to fix and stain the samples. Then study them under an oil-immersion objective to determine whether the microbes found are primarily gram-positive or gram-negative in these different habitats. Use sterilizing techniques throughout this project.

Analysis

Were the bacteria found in the hay infusion and pond water primarily gram-positive or -negative? How about those found in other habitats? Are most of the bacterial cells found in nature gram-positive or gram-negative? If they appear mixed, determine if any of the habitats contain primarily one or the other types of cells. Did the pond or your socks, for example, contain primarily gram-positive or -negative cells?

Going further

- Devise an experiment to see if common disinfectants and germicides used in the home, such as Lysol, are better suited to control gram-positive or gram-negative bacteria.
- Use a variety of other types of differential stains and agar plates to identify cell types.

Suggested research

- Research the numerous types of media and stains used to differentiate bacterial cells.
- Read about differences between gram-positive and gram-negative cells. Are either more common in pathogens or other categories of cells?

23

Electromagnetic fields & aquatic organisms

Does electromagnetic radiation affect populations of brine shrimp?

Background

Recent studies indicate that electromagnetic fields might affect the health of living organisms, including human beings. An electromagnetic field (*EMF*) is produced by almost any device with an electric current running through it, including power lines, electrical appliances, electric blankets, and personal computers.

People are not the only organisms subjected to this type of radiation. Experiments that study the effects of EMF on both terrestrial and aquatic organisms other than humans are important for two reasons: first, they could demonstrate the effect of this type of radiation on an ecosystem; second, they could indicate health risks to ourselves. If this radiation proves harmful to lower forms of life, should we be concerned about our own health?

Project overview

This experiment is designed to test the effects of an electromagnetic field on populations of brine shrimp in the lab. These small crustaceans play an important role in the marine food web. They are food for small fish and other crustaceans.

Electromagnetic fields are measured by their frequency. Normal "house current" oscillates 60 times per second, or *Hertz (Hz)*. This project simulates the actual current found passing through typical electronic devices in the home, by using a coil wrapped around the subjects (brine shrimp).

The project is divided into three steps. In the first, you set up the apparatus to simulate a typical, commercial brine-shrimp hatchery. In the second, you prepare the coils to produce an electromagnetic field around the experimental groups of young brine shrimp. In the third, you run the actual experiment and then take samples and counts of the shrimp in the experimental and control groups at various intervals. You'll use a suction vacuum to filter out the brine shrimp to make them easier to count.

Materials list

Unless noted otherwise, the equipment listed below might be available in your school. If it is not at your school, see if you can borrow the equipment from other school labs or possibly a local hospital or university. The equipment is available from a scientific supply house, but would be expensive to purchase.

- 500-gram can of brine shrimp cysts (available from a pet store)
- Package of instant ocean aquarium sea salt (from a pet store)
- Can of hatchfry (from a pet store)
- 2 three-gang air valves (from a pet store)
- 2 aquarium air pumps (from a pet store)
- 6 glass-fiber aquarium air stones (from a pet store)
- 6 50-cm lengths of aquarium tubing (from a pet store)
- 2 90-cm lengths of aquarium tubing (from a pet store)
- 6 500-ml separatory funnels
- 6 adjustable ring stands
- 2.5 liters of distilled water

- 6 disposable glass pipettes
- Spatula
- 200-ml Erlenmeyer flask and stopper
- 150-ml beaker
- 250-ml graduated cylinder
- Hydrometer
- Stirring hot plate
- Parafilm
- 30 Whatman #40 ashless filter papers (11.0 cm)
- Pipette pump
- 500-ml sidearm Erlenmeyer flask with rubber vacuum hose
- Plastic funnel with vacuum top
- Vacuum pump
- Forceps
- Hand-held counter
- Dissection microscope
- Balance
- 6-ml pipettes
- Electrical wire (copper, 14-gauge), 6 meters each of white and black (available from a hardware store)
- Porcelain lightbulb socket and 25-watt or less bulb (from a hardware store)
- Wire stripper (from a hardware store)
- UL-listed surge protector (from a hardware store)
- Three-prong electrical plug (from a hardware store)
- Roll of electrical tape (from a hardware store)
- Portable AM radio

Procedures
Preparing the apparatus

Set up six 500-ml separatory funnels in six ring stands as shown in Fig. 23-1. Position them in two sets of three with 50 cm between the two sets. Rinse each funnel with cool tap water, then fill each funnel with tap water and let stand for five minutes.

While the tap water is standing in the funnels, cut six pieces of aquarium tubing to 50 cm and two pieces to 90 cm as shown in the figure. Connect three of the 50-cm tubes to the three outlet nipples of a three-gang air valve. Repeat the procedure with the other air valve.

Attach one air stone to the end of each piece of 50-cm tubing. One air stone will be placed into each funnel. Attach one end of the 90-cm tubing to the air-outlet on an aquarium air pump and the other

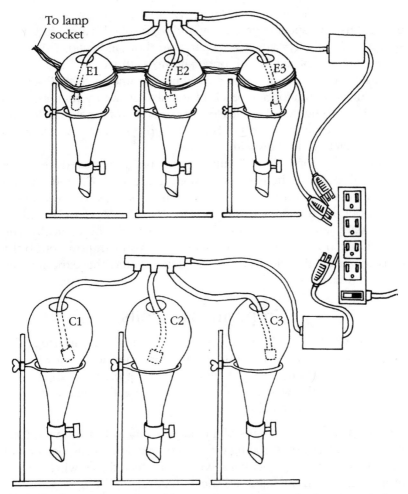

23-1 *Each group consists of three separatory funnel tubes mounted on ring stands, containing an air stone with tubing leading to the air pump. In addition, the experimental group has coils wrapped around the funnels.*

end to the inlet nipple of the three-gang valve. Repeat this with the other 90-cm tubing as shown in the figure. Pour the tap water out of the separatory funnels, then rinse with distilled water.

You will now prepare to add the salt to the funnels. Weigh 60 grams of instant ocean synthetic sea salt on a balance and pour it into a large flask. Use a stirring hot plate to dissolve the salt in 2 liters of distilled water at 24°C.

Use a hydrometer to determine the specific gravity of the solution. Adjust the specific gravity by adding distilled water until it assumes 1.020 salinity, which is 30 parts per thousand (*ppt*). This is the perfect amount of salt for the brine shrimp to thrive.

Pour the salt solution into a 2000-ml Erlenmeyer flask. Pour 250 ml of this salt solution into each of the six separatory funnels. (You might have to tighten the release valves on each funnel to halt leakage.) Measure out 0.5 grams of brine shrimp cysts and inoculate each of the separatory funnels with this amount.

Cover the tops of each funnel with parafilm to reduce water evaporation. Label the first set of three funnels *E1*, *E2*, and *E3*, and record the date and time of inoculation. (*E* stands for experimental.) Label the second set of three funnels *C1*, *C2*, and *C3*, and include date and time of inoculation. (*C* stands for control.) The funnels are now prepared and inoculated. You can now set up the coils to create the electromagnetic field around the experimental set of funnels.

Preparing the coils

⚠ ***Caution!*** The electrical apparatus created in this step must be checked by your sponsor or other adult knowledgeable about electrical circuitry. Do not plug this apparatus into the wall socket unless you are authorized to do so by your sponsor.

Prepare two 6-meter lengths of 14-gauge copper electrical wire (one white, one black) by stripping off 1 cm of insulation from one end of each wire, exposing the bare wires. Connect these wires to the proper screws on the lightbulb socket fixture.

Starting about 45 cm from the end of the white (positive) wire, wrap the wire six times around an object with a diameter of 9 cm. (This is approximately the same diameter of the separatory funnels.) Tape the coils together, and repeat this procedure for the second and third groups of coils as in Fig. 23-2. Take the first funnel (E1) out of its ring stand as seen in Fig. 23-1. Fit one group of coils around the upper portion of the funnel and then gently place the separatory funnel back in its stand. Repeat this procedure for E2 and E3. Place the bulb in the socket.

Use a wire stripper to remove 1 cm of insulation off the remaining ends of the two wires. Connect the positive wire to the positive screw of a three-prong plug and connect the negative wire to the negative screw. Replace the plug insulation (if any) and insert the plug into the surge protector.

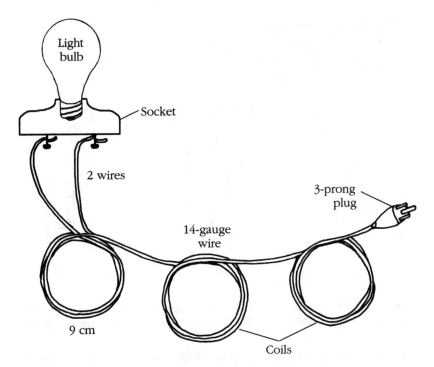

23-2 *Prepare the coils by wrapping them around an object that is 9 cm in diameter. At one end, the wires terminate in a plastic electrical cap. At the other end, they are connected into a three-prong plug.*

Turn on the power and test for the presence of an electromagnetic field within the coils with a portable AM radio. You'll get static if the field exists. Maintain a constant room temperature of 24°C and a constant light-and-dark cycle (preferably nine hours of light and 15 hours of dark) throughout the project.

Allow the populations to stabilize for 24 hours after inoculation. Remove the air stones and replace them with 1-ml plastic pipettes for a gentler airflow. You are now ready to perform the first population counts.

Performing population counts

Prepare the vacuum apparatus as shown in Fig. 23-3 by connecting one end of a rubber vacuum hose to the vacuum supply. Connect the other end to the outlet of a 500-ml sidearm flask as shown. Attach the plastic funnel and vacuum cover unit.

Cut six Whatman #40 ashless filter discs (or papers) to fit inside a funnel's vacuum cover unit. Mark the papers into sections as shown

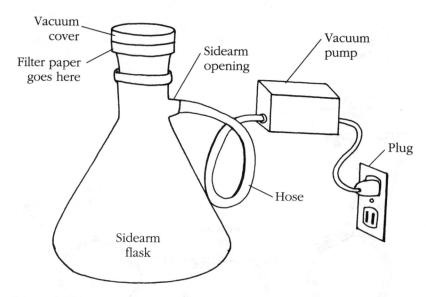

23-3 *The filter paper is placed in the vacuum cover. The vacuum pump creates a vacuum that sucks the liquid through the filter into the flask.*

in Fig. 23-4 to make counting easier. With a pencil, draw a circle 1.5 cm in diameter inside the middle of each filter disc. Then draw four lines radiating out in an *X* pattern from the center circle, as shown in the figure. Label these sections of each disk *1* through *5*. Also, label each disk to match a funnel: *E1, E2, E3, C1, C2,* and *C3*. Wet the disks slightly, starting with E1, and put this disk in the vacuum cover unit. Turn the vacuum supply on to half power.

Insert a long pipette with a pipette bulb in the E1 funnel to the 12 cm mark of the pipette and draw 1 ml of culture solution. Release the

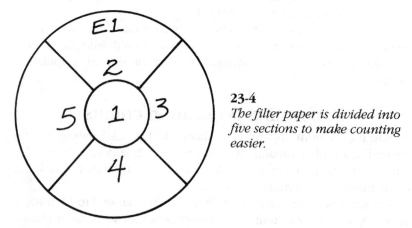

23-4
The filter paper is divided into five sections to make counting easier.

solution onto the first section on the filter paper. Repeat this procedure for all five sections on the paper. View the dried filter paper
discs under a dissecting microscope at low power (100X). Use a
hand-held counter to count the number of visible shrimp in each section and chart these populations.

Repeat this procedure for funnels E2, E3, C1, C2 and C3, repeat-
ing the draw-and-release procedure five times for each separatory
funnel.

Repeat the sampling and counting technique described above
every 24 hours through 144 hours. (The brine shrimp should be fed
at 48 hours, at a feeding rate of 0.1 gram of diet per separatory funnel. After 96 hours, feed at a rate of 0.02 grams into each of the six
separatory funnels.)

Analysis

Compare the population counts of the control funnels with the experimental funnels over time. Are there differences? Did they occur
after the first count (24 hours), or did it take longer to see differences?
Does it appear that the EMR had an effect on the health and survival
of these organisms?

Going further

This project uses one level of radiation. Devise a similar experiment
that could manipulate the degree of radiation reaching the subjects.
This can easily be done by adjusting the distance of the subjects from
the coils. Rent a gaussmeter to take radiation readings.

Suggested research

- Research the directionality of EMF. Is the angle of the coils
  used in this project important to the results?
- Research whether any studies have been performed on the
  effect of EMF on aquatic organisms.

Appendix A

Using metrics

Most science fairs require that all measurements be taken using the metric system as opposed to English units. Meters and grams, which are based on powers of 10, are far easier to use during your experimentation than feet and pounds.

You can convert English units into metric units if you need to, but it is easier to simply begin with metric units. If you are using school equipment such as flasks or cylinders, check the graduations to see if any have metric units. If you are purchasing your glassware (or plasticware), be sure to order metric graduations.

Conversions from English units to metric units are given below, along with their abbreviations as used in this book. (All conversions are approximate.)

Length

1 inch (in) = 2.54 centimeters (cm)
1 foot (ft) = 30 cm
1 yard (yd) = 0.90 meters (m)
1 mile (mi) = 1.6 kilometers (km)

Volume

1 teaspoon (tsp) = 5 milliliters (ml)
1 tablespoon (Tbsp) = 15 ml
1 fluid ounce (fl oz) = 30 ml
1 cup (c) = 0.24 liters (l)
1 pint (pt) = 0.47 l
1 quart (qt) = 0.95 l
1 gallon (gal) = 3.80 l

Mass

1 ounce (oz) = 28 grams (g)
1 pound (lb) = 0.45 kilograms (kg)

Temperature

32° Fahrenheit (F) = 0° Celsius (C)
212°F = 100°C (See Fig. A-1)

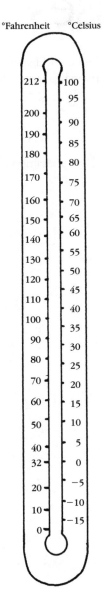

A-1 *Use this thermometer to convert Celsius to Fahrenheit and vice versa.*

Appendix B

Addresses

For information about the International Science and Engineering Fairs and valuable information about adult sponsorship, write or call:

The Science Service, Inc.
1719 N Street NW
Washington, DC 20036
(202) 785-2255

You can order equipment, supplies, and live specimens for projects in this book from these companies:

Ward's Natural Science Establishment, Inc.
5100 West Henrietta Road
Rochester, New York 14692
(800) 962-2660
or
815 Fiero Lane
P.O. Box 5010
San Luis Obispo, California 93403
(800) 872-7289

Blue Spruce Biological Supply Company
221 South Street
Castle Rock, Colorado 80104
(800) 621-8385

The Carolina Biological Supply Company
2700 York Road
Burlington, North Carolina 27215
Eastern U.S.: (800) 334-5551
Western U.S.: (800) 547-1733

Connecticut Valley Biological
82 Valley Road
P.O. Box 326
Southampton, Massachusetts 01073
(413) 527-4030

Fisher Scientific
4901 West LeMoyne Street
Chicago, Illinois 60651
(800) 955-1177

Frey Scientific Company
905 Hickory Lane
P.O. Box 8101
Mansfield, Ohio 44901
(800) 225-FREY

Nasco
901 Janesville Avenue
P.O. Box 901
Fort Atkinson, Wisconsin 53538
(800) 558-9595

Nebraska Scientific
3823 Leavenworth Street
Omaha, Nebraska 68105
(800) 228-7117

Powell Laboratories Division
19355 McLoughlin Boulevard
Gladstone, Oregon 97027
(800) 547-1733

Sargent-Welch Scientific Company
P.O. Box 1026
Skokie, Illinois 60076
(312) 677-0600

Southern Biological Supply Company
P.O. Box 368
McKenzie, Tennessee 38201
(800) 748-8735

Glossary

abstract A brief written overview that describes your project, usually less than 250 words and often required at science fairs.

aerobe An organism that requires oxygen to survive.

agar Gelatin-like substance produced from red seaweed, used as a solid support for microbial cultures.

algae Simple, aquatic, photosynthetic plants. They may be one-celled, microscopic organisms or live in large macroscopic colonies. They are found in almost all aquatic habitats.

amoeba A simple, one-celled protozoan that moves by creating pseudopodia.

anaerobe Organisms that can live in the absence of oxygen.

antimicrobial A substance that inhibits the growth of microbes using biological or chemical means.

antiseptic A substance that prevents decay by preventing the growth of microbes.

autoclave A device that sterilizes objects by producing steam under pressure.

autotroph An organism that produces its own food (by photo-synthesis using the sun or from inorganic chemical energy). Also called a *producer.*

backboard The vertical, self-supporting panel used in a science fair display. The board usually displays the problem, hypothesis, photos of the apparatus used, organisms, and other important aspects of the project, as well as analyzed data in the form of charts and tables. Most fairs have size limitations for backboards.

bacteria This large group of single-celled, microscopic organisms has no nuclear membrane (*procaryotic*), no chlorophyll, and usually reproduces by simple fission. They grow in colonies visible to the unaided eye.

biochemistry The study of chemical processes in living organisms.

biocontrol The use of organisms to control pests, also called *biological control.*

biofilm A thin layer of microbial growth on a surface; occurs in a moist or aquatic habitat.

buffer Any substance in a liquid that tends to resist changes in the pH level.

cell The basic unit of life. Cells are bags containing a liquid interior (the *cytoplasm*). Enclosed are *organelles* and genetic material. The bag itself is the *cell membrane.* Organisms are either *unicellular* (such as protists) or multicellular (such as humans).

colony A visible population of cells growing on a solid medium such as agar.

consumer An organism that must consume (eat) its food as opposed to producers, which make their food. Also called *heterotrophs.*

control group A test group in an experiment that provides a baseline for comparison, where no experimental factors or stimulus are introduced.

culture A population of microbes grown in a medium such as nutrient agar under controlled conditions such as in an incubator.

dependent variable The variable in an experiment to be measured. For example, if you are testing the death rate of microbes living in the soil after exposure to a pesticide, the death rate is the dependent variable and the pesticide is the independent variable.

display Refers to the entire science fair exhibit, of which the backboard is a part.

experimental variable Also called the *independent variable,* refers to the aspect or factor of an experiment that is manipulated or changed for comparison. For example, the amount of a fungicide used to kill molds is an experimental variable.

eukaryotic A cell containing internal organelles and a nucleus.

experimental group A test group that is subjected to experimental factors or stimulus for the sake of comparison with the control group. The experimental group is the one exposed to the factor being tested, such as a microbial culture exposed to a variety of fungicides.

filter sterilization A method of sterilizing a liquid by forcing the liquid through a filter with pores less than 0.20 microns; the liquid is sterilized because the microbes cannot fit through the small pores.

food chain A simple representation of "who eats what" in an ecosystem; represented by one-to-one relationships.

fungus Primitive plants incapable of photosynthesis. Most are *saprophytic,* meaning they feed on decaying organic matter.

habitat Refers to the place where an organism lives, such as an aquatic or terrestrial habitat.

hemolytic Having the ability to breakdown (*lyse*) blood cells.

heterotroph Organisms that require an external source of organic chemical energy (food) to survive (as opposed to autotrophic organisms). Also called *consumers*.

host The organism which supports the life of a parasite; for example, humans are the host for cold viruses.

hyphae The filaments of a mold or a mushroom; in molds, the tip of the hyphae produce spores that produce the next generation.

hypothesis An educated guess, formulated after thorough research, to be shown true or false through experimentation.

infection A growth of microorganisms within a host causing illness in the host.

inoculum The starting material for a microbial culture.

inorganic matter Refers to substances that are not alive and did not come from decomposed organisms.

invertebrates Organisms with no backbones such as insects, starfish, and lobsters.

journal Also called the *project notebook*, contains all notes on all aspects of a science fair project from start to finish.

lyse To break open, usually referring to cells.

macroscopic Large enough to see with the unaided eye.

metabolism The sum of the physical and biochemical reactions necessary for life.

microbe An organism visible only with a microscope.

mold A general term for many kinds of simple, filamentous fungi.

morphology The study of the appearance of an organism, including its shape, texture, and color.

mycelium The main body of a multicellular fungi.

mycoses Infectious diseases caused by fungi.

nematodes Also called *roundworms*; small, unsegmented, microscopic worms found in most habitats in great numbers. Most are harmless, but a few are parasitic.

nucleus A membrane-enclosed structure that contains genetic material in a eukaryotic cell.

obligate pathogens Microorganisms that must have a host to survive and reproduce. In comparison to *facultative pathogens*, which are accidental contaminants of a host and can survive outside of a host.

observations A form of qualitative data collection.

organelle A membrane-enclosed structure within a cell in eukaryotic organisms.

organic Refers to substances that compose living or dead, decaying organisms and their waste products. Carbon is the primary element.

parasite An organism that lives in or on one or more organisms (hosts) during a portion of its life. The host is usually not killed in the process.

pathogens Organisms that cause disease in other organisms.

petri dish A shallow plate used to hold a culture medium such as nutrient agar.

pipette A long, thin tube that holds a known amount of liquid; used to transfer liquids.

predator An animal (consumer) that eats other animals for nourishment.

producer An organism that makes its own chemical energy (food) usually using energy from the sun.

protists Members of a kingdom of living things, composed of single-celled eukaryotes that do not have a cell wall. Some have chlorophyll while others do not.

protozoa A group of protists that do not contain chlorophyll. Complex, single-celled animals (*eukaryotes*).

qualitative studies Experimentation where data collection involves observations but no numerical results.

quantitative studies Experimentation where data collection involves measurements and numerical results.

raw data Any data collected during the course of an experiment that has not been manipulated in any way, as opposed to *smooth data.*

research Also called a *literature search*, refers to locating and studying as much of the existing information about a subject as possible.

resolving power, microscope The smallest distance between two objects in which the two objects can still be distinguished from one another. If the two objects are beyond the resolving power of a microscope, they appear as one.

scavenger An organism that consumes dead organic matter.

scientific method The basic methodology of all scientific experimentation including: 1) the statement of a problem to be solved or question to be answered through further science, 2) the formulation of a hypothesis and 3) experimentation to determine if the hypothesis is true or false. Experimentation includes data collection, analysis, and arriving at a conclusion.

selective media Media that contains one or more chemical agents that do not permit some microbes to grow, but do permit the growth of the microbe of interest. Also called *differential media.*

smooth data Raw data that has been manipulated to provide understandable information, such as graphs and charts that represent totals, averages, and other numerical analysis.

species Organisms with the potential to breed and produce viable offspring.

spore-forming bacteria Prokaryotes that produce spores which are resistant to drying and can survive difficult environmental conditions.

statistics Refers to analyzing numerical data.

sterile The absence of all life.

stimulus An "event" that prompts a reaction or a response.

variables A factor that is changed to test the hypothesis.

virus A package of genetic material surrounded by a protein capsule that requires a living host to reproduce.

Bibliography

Suggested reading

For more information about microbiology, try any of these books:

Anderson, D.A. and R.J. Sobieski. 1980. *Introduction to Microbiology, 2nd ed.* St. Louis: The C.V. Mosby Company.

Bleifeld, Maurice. 1988. *Experimenting with a Microscope.* New York: Franklin Watts.

Jahn, T.L., E.C. Bovee and F.F. Jahn. 1979. *How to Know the Protozoa, 2nd ed.* Dubuque, Iowa: Wm. C. Brown Co. Publ.

Nardo, Don. 1991. *Germs: Mysterious Microorganisms.* San Diego: Lucent Books.

Oxlade, C. and C. Stockley. 1989. *The World of a Microscope.* Tulsa: EDC Publishers.

Pelczar, M.J. and Roger Reid. 1965. *Microbiology, 2nd ed.* New York: McGraw-Hill.

Poindexter, Jeanne. 1971. *An Introduction to Protists.* New York: MacMillan.

Prescott, G.W. 1978. *How to Know the Freshwater Algae, 3rd ed.* Dubuque, Iowa: Wm. C. Brown Co. Publ.

Sabin, Francene. 1985. *Microbes and Bacteria.* Mawhwah, NJ: Troll Associates.

Stewart, Gail. 1992. *Microscopes: Bringing the Unseen World Into Focus.* San Diego: Lucent Books.

Tortora, Gerald. 1989. *Microbiology: An Introduction.* Redwood City, CA: Benj. Cummings Pub.

If you are new to science fairs, the following are a few good books to read. They cover all aspects of entering a science fair.

Bombaugh, Ruth. 1990. *Science Fair Success.* Hillside, NJ: Enslow Publishers.

Irtz, Maxine. 1987. *Science Fair: Developing a Successful and Fun Project.* Blue Ridge Summit, PA: TAB Books.

Tocci, Salvatore. 1986. *How to Do a Science Fair Project.* New York: Franklin Watts.

The following books can all be used for additional science-fair project ideas. Although not specifically about microbiology, many involve microbes or can be adapted to create projects about microbes.

Bochinski, Julianne. 1991. *The Complete Handbook of Science Fair Projects.* New York: Wiley & Sons, Inc.

Dashefsky, H. Steven. 1994. *Entomology: High School Science Fair Experiments.* Blue Ridge Summit, PA: TAB/McGraw-Hill.

Dashefsky, H. Steven. 1994. *Environmental Science: High School Science Fair Experiments.* Blue Ridge Summit, PA: TAB/McGraw-Hill.

Dashefsky, H. Steven. 1994. *Zoology: High School Science Fair Experiments.* Blue Ridge Summit, PA: TAB/McGraw-Hill.

Gardner, Robert. 1989. *More Ideas for Science Fair Projects.* New York: Franklin Watts.

Irtz, Maxine. 1991. *Blue-Ribbon Science Fair Projects.* Blue Ridge Summit, PA: TAB Books.

VanCleave, Janice. 1990. *Biology for Every Kid: 101 Easy Experiments that Really Work.* New York: John Wiley & Sons, Inc.

VanCleave, Janice. 1993. *A+ Projects in Biology.* New York: John Wiley & Sons, Inc.

Witherspoon, James D. 1993. *From Field to Lab: 200 Life Science Experiments for the Amateur Biologist.* Blue Ridge Summit, PA: TAB Books.

Index

A
adult supervision, x-xii
agar, 4
agar slants, 5
air conditioner and mold presence,
  93-98, **94**, **96**, **98**
algae, 3, **4**
Analysis section of projects, xiv, xv
analysis, scientific method, 22
antimicrobial products, 109-122
  biofilms and biofouling, 118-122,
  119, **121**
  household disinfectants vs. resistant
  bacteria, 111-117
  paint, 109, 118-122, **119**, **121**
aseptic techniques (*see* sterile
  techniques)
autoclave, 7, **7**
autotrophs, 3

B
Background section of projects, xiii,
  xiv, 24
bacteria, 3, **4**
Baermann Funnel Extraction
  technique, 133-134, **133**
bathroom cleanliness, 123-127, **125**
before beginning a project, 26
binocular scope, 13
biofilms and biofouling, 118-122,
  119, **121**
biofouling, 118-122, **119**, **121**
bottled water and microbes, 87-92,
  90, **91**
buffers and microbes, 129, 137-140,
  138
building on past projects, 23
bunsen burner to sterilize, 8-9, **9**

C
cell counts, 18-19, **18**
colonies, 4
compound microscopes, 12-13, **13**,
  15-17
conclusions, scientific method, 22-23
controlling and using microbes, 31-48
  fungal control of insect pests, 38-43,
  41
  oysters, hot sauce, and bacteria, 33-
  37, **35**, **36**
  tree bark as bacterial inhibitor, 44-
  48, **46**, **47**
culture media, 5
cultures, 4

D
differential stains, 141-145, **143**, **144**
Discover, 25
disease-causing organisms, xi-xii
displays, 29
dissecting scopes, 15, **16**

E
electromagnetic radiation vs.
  microbes, 129
  aquatic life, 146-153, **149**, **151**, **152**
  soil microbes, 131-6, **133**, **135**
experimentation, scientific method,
  21-22

F
food
  buffers and microbes, 129, 137-140,
  138
  garlic, health, microbes, 67-84
  hot sauce, and bacteria, 33-37, **35**, **36**

Boldface numbers indicate illustrations.

food, *cont.*
  oysters, hot sauce, and bacteria, 33-37, **35**, **36**
  raw food and bacterial infection, 33-37, **35**, **36**
  spices vs. microbes, 69-72, **71**, **72**
  useful microbes, 129
fungal control of insect pests, 38-43, **41**
fungi, 3, **4**

G

garlic, health, microbes, 67-84, **80**
  garlic juice vs. microbial growth, 73-78, **75**, **76**, **77**
  heating garlic and bacterial control, 79-84, **82**, **83**
  spices vs. microbes, 69-72, **71**, **72**
glassware sterilization, 7-8, **7**
Going Further section of projects, xiv, xv, 24
gram stains, 141-145, **143**, **144**
growth media, 4, 5

H

handshakes and spread of bacteria, 55-60, **57**, **59**
hemocytometer, 18
heterotrophs, 3
home and microbial presence, 123-127
  air conditioner and mold presence, 93-98, **94**, **96**, **98**
  bathroom cleanliness, 123-127, **125**
  kitchen cleanliness, 123-127, **126**
  household disinfectants vs. resistant bacteria, 111-117, **114**, **115**, **117**
  mold presence, 93-98, **94**, **96**, **98**
  resistant bacteria, 111-117, **114**, **115**, **117**
  vacuum cleaner and mold presence, 93-98, **94**, **96**, **98**
hot sauce, and bacteria, 33-37, **35**, **36**
household disinfectants vs. resistant bacteria, 111-117, **114**, **115**, **117**
human body and microbe presence, 99-103, **101**, **102**
hypothesis, scientific method, 21

I

in vitro vs. in vivo, 23
incubation, 17-19
  cell counts, 18-19, **18**
  counting cultures, 116
  diluting cultures, 115-116
  hemocytometer, 18
  incubator, 17-18, **17**
  isolating cells, 114-115, **115**
  ocular micrometers, 18-19, **18**
  plate preparation, 113
  readings, 113-114
  selecting and culturing cells, 114
  streaking, 114-115, **115**
  vernier calipers, 19, **19**
  zone of inhibition, 19
  zones on petri dish, **114**
inoculating loops, 8-9, **8**, **10**
inoculation, 6
insects
  fungal control of insect pests, 38-43, **41**
  insect transport of microbes, 51-54, **52**, **53**
International Science and Engineering Fairs
  information, 156-157
  rules, x-xi
interviewing specialists, 25-26

J
judging, 29

K
kitchen cleanliness, 123-127, **126**

L
literature search, 27

M
magazines as information sources, 25
magnification power, 13
Materials Lists, xiii, xiv
McFarland standard, 76
mechanical stage, microscopes, 14, **14**
media sterilization, 7-8, **7**
medicines
  human body and microbe presence, 99-103, **101**, **102**

tree bark as bacterial inhibitor, 44-48, **46**, **47**
metrics, 154-155
microbes, 3, **4**
microbiology basics, 3-19
  agar, 4
  agar slants, 5
  algae, 3, **4**
  aseptic techniques (*see* sterile techniques), 5
  autotrophs, 3
  bacteria, 3, **4**
  cell counts, 18-19, **18**
  colonies, 4
  culture media, 5
  cultures, 4
  fungi, 3, **4**
  growth media, 4, 5
  hemocytometer, 18
  heterotrophs, 3
  incubation, 17-19
  microbes, 3, **4**
  microorganisms, 3
  microscopes, 12-17
  ocular micrometers, 18-19, **18**
  plates, 5
  protists, 3, **4**
  pure cultures, 4
  quantitative plating method, 104-105
  saprophytes, 3
  sources of microbes, 5
  sterile technique, 5-12
  vernier calipers, 19, **19**
  viruses, 3, **4**
microorganisms, 3
microscopes, 12-17
  binocular scope, 13
  compound microscopes, 12-13, **13**, 15-17
  dissecting scopes, 15, **16**
  light path through microscopes, 14, **15**
  magnification power, 13
  mechanical stage, 14, **14**
  monocular scope, 13
  objective lens, 13
  ocular micrometers, 18-19, **18**
  oculars, 13
  resolving power, 12
  stages, 13-14

stereoscopes, 15
substage, 14
mold, 93, **94**
  in air conditioner, vacuum cleaner, home, 93-98, **94**, **96**, **98**
monocular scope, 13
movement of microbes, 49-65
  insect transport of microbes, 51-54, **52**, **53**
  shaking hands and spread of bacteria, 55-60, **57**, **59**
  wind transport of microbes, 61-65, **63**

O
objective lens, 13
ocular micrometers, 18-19, **18**
oculars, microscopes, 13
Omni, 25
Overviews of projects, 24
oysters, hot sauce, and bacteria, 33-37, **35**, **36**

P
paint as antimicrobial products, 109, 118-122, **119**, **121**
pathogens, xi-xii, 6
pipettes, 11-12, **11**, **12**
plates, 5
Popular Science, 25
presence of microbes, 85-108
  bottled water and microbes, 87-92, **90**, **91**
  human body and microbe presence, 99-103, **101**, **102**
  mold in a/c, vacuum, home, 93-98, **94**, **96**, **98**
  soil microbes, 104-108, **106**
problem statement, scientific method, 21
Procedures section of projects, xiii, xv
Project Overviews, xiii, xiv
projects, xiii
protists, 3, **4**
pure cultures, 4

Q
quantitative plating method, 104-105

R

Reader's Guide to Periodical Literature, 25
replication, 22
research (*see* scientific research)
research papers, 28-29
research plan, 28
resistant bacteria, 111-117, **114**, **115**, **117**
resolving power, 12

S

safety, x-xii
saprophytes, 3
scheduling the experiment, 26-27
science fair guidelines, 28
scientific method, 21-23
  analysis, 22
  conclusions, 22-23
  experimentation, 21-22
  hypothesis, 21
  in vitro vs. in vivo, 23
  problem statement, 21
  replication, 22
scientific research, 20-23
  Background section of projects, 24
  before beginning a project, 26
  building on past projects, 23
  displays, 29
  Going Further section of project, 24
  in vitro vs. in vivo, 23
  interviewing specialists, 25-26
  judging, 29
  literature search, 27
  magazines as information sources, 25
  Overviews of projects, 24
  replication, 22
  research papers, 28-29
  research plan, 28
  scheduling the experiment, 26-27
  science fair guidelines, 28
  scientific method, 21-23
  sources of help and information, 25
  specialists in the field, 25-26
  Suggested Research section of project, 24
  television programs as information sources, 25
shaking hands and spread of bacteria, 55-60, **57**, **59**

soil microbes, 104-108, **106**
  electromagnetic radiation vs. microbes, 131-136, **133**, **135**
sources of help and information, 25
sources of microbes, 5
sources of supplies and information, 25, 156-157
specialists in the field, 25-26
spices vs. microbes, 69-72, **71**, **72**
spreading rods, 10-11, **10**
stages, microscopes, 13-14
stereoscopes, 15
sterile techniques, 5-12
  autoclave, 7, **7**
  bunsen burner to sterilize, 8-9, **9**
  glassware sterilization, 7-8, **7**
  inoculating loops, 8-9, **8**, **10**
  inoculation, 6
  media sterilization, 7-8, **7**
  pathogens, 6
  pipettes, 11-12, **11**, **12**
  spreading rods, 10-11, **10**
substage, microscopes, 14
Suggested Research section of projects, xiv, xvi, 24
survey of common bacteria, 141-145, **143**, **144**

T

television programs as information sources, 25
tree bark as bacterial inhibitor, 44-48, **46**, **47**
turbidity, 76, **140**

V

vacuum cleaner and mold presence, 93-98, **94**, **96**, **98**
vernier calipers, 19, **19**
viruses, 3, **4**

W

water, bottled water and microbes, 87-92, **90**, **91**
weather, wind transport of microbes, 61-65, **63**
wind transport of microbes, 61-65, **63**

Z

zone of inhibition, 19

About the author

Steven Dashefsky is an adjunct professor of environmental science at Marymount College in Tarrytown, New York. He is the founder of the Center for Environmental Literacy, which was created to educate the public and business community about environmental problems and solutions. He holds a B.S. in biology and an M.S. in entomology, and is the author of over 10 books that simplify science and technology.